Recent Advances in Catalytic Surfaces and Interfaces

Recent Advances in Catalytic Surfaces and Interfaces

Guest Editors

Michalis Konsolakis
Irene Groot

MDPI

Basel • Beijing • Wuhan • Barcelona • Belgrade • Novi Sad • Cluj • Manchester

Guest Editors
Michalis Konsolakis
School of Production Engineering and Management
Technical University of Crete
Chania
Greece

Irene Groot
Leiden Institute of Chemistry
Leiden University
Leiden
The Netherlands

Editorial Office
MDPI AG
Grosspeteranlage 5
4052 Basel, Switzerland

This is a reprint of the Special Issue, published open access by the journal *Surfaces* (ISSN 2571-9637), freely accessible at: https://www.mdpi.com/journal/surfaces/special_issues/CEPOHJBBH9.

For citation purposes, cite each article independently as indicated on the article page online and as indicated below:

Lastname, A.A.; Lastname, B.B. Article Title. *Journal Name* **Year**, *Volume Number*, Page Range.

ISBN 978-3-7258-5349-6 (Hbk)
ISBN 978-3-7258-5350-2 (PDF)
https://doi.org/10.3390/books978-3-7258-5350-2

Contents

About the Editors

Michalis Konsolakis

Michalis Konsolakis is currently a Full Professor of "Heterogeneous Catalysis & Surface Science" at the School of Production Engineering & Management of the Technical University of Crete, Greece. His research activities are mainly focused on the areas of heterogeneous catalysis and surface science, with a particular emphasis on structure–property relationships. Recently, he has mainly focused on the rational design and nano-engineering of metal oxide catalysts by means of advanced synthetic and promotional routes. His published work includes >250 articles in international peer-review journals and conference proceedings. He is a member of the Editorial Board of several international journals in the fields of materials and surface science, serving also as a regular reviewer in >100 scientific journals and research funding agencies.

Irene Groot

Irene Groot is Professor of Surface and Interface Science at Leiden University. Her main research interests include facilitating an atomic-scale understanding of heterogeneous catalysis and the growth of two-dimensional materials. She focuses on investigating these processes as they occur under industrially relevant conditions. To achieve this, she has developed novel equipment and measurement techniques to investigate catalysts at work. To mitigate the societally relevant problems of air pollution and climate change, catalysts are of vital importance. New and better catalysts are needed to assist in the transition from fossil fuels to sustainable energy and to enable the use of CO_2 as a feedstock to produce high-value chemicals and fuels. Therefore, topics of interest for Groot include Fischer–Tropsch synthesis, hydrodesulfurization, and nanoplastics. In the field of 2D materials, her interests lie in their growth on liquid metal catalysts and in doping graphene with h-BN. In these projects, Groot has collaborated with partners from both academia and industry.

Preface

Surface and interface engineering are among the most efficient strategies to develop innovative and advanced catalytic materials. A prerequisite for this is a fundamental understanding of the structure–performance relationships at the (near-) atomic scale; these, however, remain a formidable challenge due to the complexity of heterogeneous catalytic processes. Recent progress in nanosynthesis with uniform and well-defined structures, fine-tuning engineering strategies (size/shape control), advanced characterization techniques, and theoretical calculations have offered unique opportunities to improve the fundamental understanding of surface and interface phenomena, which in turn could pave the way for the rational design of novel electro-catalytic systems. Herein, this Reprint involves some comprehensive studies on surface and interface functionalization in electro-catalysis, with an emphasis on the use of advanced synthesis/modification routes and characterization techniques, aiming to unravel the role of surface and interface sites and the establishment of reliable structure–property relationships.

Michalis Konsolakis and Irene Groot
Guest Editors

surfaces

Article

Impact of Impregnation pH on NiMo Surface Species in Al_2O_3-Supported Catalysts for Green Diesel Production

Alexis G. Mijares-Alvarado [1], Acela López-Benítez [2,*], Rebeca Silva-Rodrigo [3], José A. Rodríguez-Ávila [1] and Alfredo Guevara-Lara [1,*]

1 Área Académica de Química, Instituto de Ciencias Básicas e Ingeniería, Universidad Autónoma del Estado de Hidalgo (UAEH), Mineral de la Reforma 42184, Hidalgo, Mexico; mi429388@uaeh.edu.mx (A.G.M.-A.); josear@uaeh.edu.mx (J.A.R.-Á.)

2 Department of Nanomaterials, Unidad Profesional Interdisciplinaria de Ingeniería Campus Hidalgo (UPIIH), Instituto Politécnico Nacional (IPN), San Agustín Tlaxiaca 42162, Hidalgo, Mexico

3 Centro de Investigación en Petroquímica, Tecnológico Nacional de México, Instituto Tecnológico de Ciudad Madero, Ciudad Madero 89600, Tamaulipas, Mexico; rebeca.sr@cdmadero.tecnm.mx

* Correspondence: aclopezb@ipn.mx (A.L.-B.); guevaraa@uaeh.edu.mx (A.G.-L.)

Abstract: Green diesel is a high-quality biofuel obtained through the transformation of triglycerides into linear alkanes. In order to obtain green diesel, this study investigates the impact of impregnation pH on the surface species of $NiMo/Al_2O_3$ catalysts in the hydroprocessing of soybean oil. NiMo catalysts supported on Al_2O_3 were synthesized at different pH values (pH = 7 and 9). In the oxide state, solids were characterized by UV-Vis diffuse reflectance, Raman, and FT-IR spectroscopies, and, in the sulfide state, they were characterized by HR-TEM. The results show that the pH of impregnation significantly determines the surface species formed. An impregnation at pH = 7 favors the formation of $Ni^{2+}{}_{(Oh)}$ and $Ni^{2+}{}_{(Oh\text{-}dis)}$ interacting with non-crystalline molybdenum trioxide, while the formation of Ni^{2+}/Al_2O_3, $Ni_{2+(Oh\text{-}dis)}$, and MoO_3 species is favored at pH = 9. These surface species play a fundamental role in the hydrogenolysis and deoxygenation steps. Catalyst impregnated at pH = 7 shows higher activity due to the formation of shorter MoS_2 slabs. This study emphasized the importance of controlling impregnation conditions for optimizing catalyst performance.

Keywords: green diesel; surface species; biodiesel; hydroprocessing; supported catalysts

1. Introduction

Biofuels represent a promising solution to meet the increasing global energy demand [1]. Recently, biofuels derived from biomass have gained significant attention, as biomass is an energy source capable of replacing fossil fuels [2,3]. Additionally, biomass is a renewable and sustainable resource [4]. Green diesel has gained recognition as one of the most advantageous biofuels because of its higher oxidative stability, high cetane number, energy density, and reduced corrosiveness [5,6]. Green diesel is a mixture of hydrocarbons that is chemically identical to petroleum diesel [7,8]. It is considered a high-value-added product because it can exhibit similar, or even superior, properties compared to fossil fuels [9,10].

Biofuel is typically produced through the transesterification of vegetable oils [11]. However, this process generates large amounts of glycerol, which can impact the quality of the final product [12]. The hydroprocessing of vegetable oils (HVO) can convert the oils into linear chain alkanes, similar to those found in diesel fuels [13,14]. This process is carried out using heterogeneous catalysts (usually NiMo or CoMo), in a hydroprocessing reactor, at high hydrogen pressures (30–80 bars) and temperatures (320–410 °C) [15,16]. The main goal of HVO is to convert triglycerides into products compatible with fuel supply and storage systems. HVO proceeds through two consecutive reaction mechanisms shown in Figure 1: hydrogenolysis (k_1), which transforms triglycerides into fatty acids (e.g., palmitic linolenic, linoleic, and oleic acids), and deoxygenation (k_2), which converts these fatty acids into linear hydrocarbons such as pentadecane, hexadecane, heptadecane, and octadecane (C_{15}–C_{18}) [17].

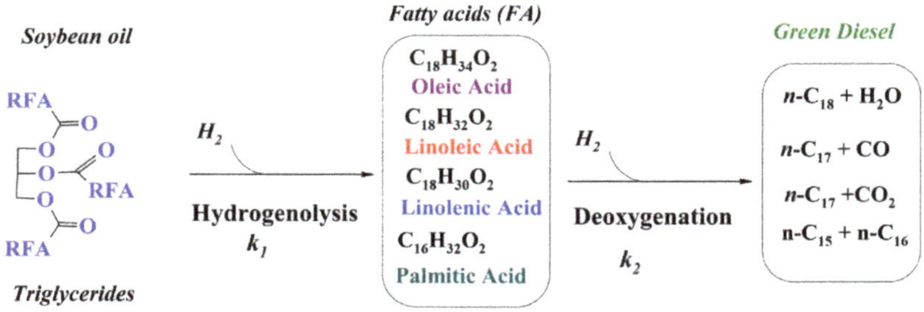

Figure 1. Mechanism of hydroprocessing of vegetable oil.

Alumina is a material widely employed as a support for catalysts because of its textural properties, high thermal stability, and moderate Lewis acidity. Catalysts such as NiMo/γ-Al_2O_3 and CoMo/γ-Al_2O_3 have been used in HVO reactions to obtain green diesel [12]. Green diesel has been obtained using oils from palm [18], castor [19], colza [20], jatropha [21], sunflower [12], and soybean [22]. In this regard, high-quality diesel can be produced from sunflower oil using a NiMoS/γ-Al_2O_3 catalyst, as mentioned by Huber et al. [12]. On the other hand, Liu et al. [23] reported that sulfurized NiMo catalysts supported on SiO_2-Al_2O_3 can produce pentadecane, hexadecane, heptadecane, and octadecane from Jatropha oil. However, the quality of the resulting green diesel depends on the type of vegetable oil used and the active phases formed on the catalyst [24]. Therefore, the development of more active catalysts able to produce high-quality green diesel is necessary.

The development of more active catalysts largely depends on the precise control of parameters during the catalyst preparation steps. The main steps include (1) preparation of the impregnation solution; (2) impregnation, where species are deposited onto the support; (3) drying and calcination; and (4) sulfurization [25]. In NiMo catalysts, the structure of the Ni-Mo oxide precursors determines the formation of active or nonactive surface species (such as $NiAl_2O_4$) [26,27]. For this reason, proper interaction between superficial Ni and Mo needs to be maintained in each step of catalyst preparation. This interaction can be influenced by factors such as additives, salts, and the pH of impregnation [28,29]. The pH plays a critical role in determining the type of Ni-Mo species deposited on the surface [30]. Nickel nitrate and ammonium heptamolybdate are commonly used to prepare the conventional impregnation solution, where $Mo_7O_{24}^{6-}/Ni^{2+}_{(Oh)}$ interaction predominates at pH = 7, while the formation of $MoO_4^{2-}/Ni^{2+}_{(Oh)}$ interaction is observed at pH = 9 [31]. These different pH values enable the study of the different surface Ni-Mo interactions. The surface species can be further modified after their deposition onto the support and during the drying, calcination, and sulfidation steps.

In this work, the effect of impregnation pH on the Ni and Mo species deposited on the surface was studied by synthesizing NiMo catalysts supported on Al_2O_3 sol–gel at pH = 7 and 9. The solids were characterized after the drying, calcination, and sulfidation steps to identify the surface species deposited on the support. The as-prepared catalysts were evaluated in a hydrotreating reactor in the hydroprocessing of soybean oil. Finally, the hydroprocessing results were correlated with the species formed as a function of the impregnation pH.

2. Materials and Methods

2.1. Synthesis of the Support (SG)

The Al_2O_3 support was obtained through the sol–gel process. To achieve this, 0.1 mol of aluminum isopropoxide (Sigma-Aldrich, >98%, St. Louis, MO, USA) was added to 0.18 L of 1-propanol (Sigma-Aldrich, >99.5%, St Louis, MO, USA) and stirred until dissolved. Then, 20 mL of deionized water was added to promote the hydrolysis and obtain the xerogel. The xerogel was thermally treated at 393 K for 720 min and then at 823 K for 300 min (5 K min^{-1}). This support is referred to as SG.

2.2. Synthesis of the NiMo Catalyst Supported on SG

Catalysts were synthesized applying the co-impregnation technique, employing an aqueous solution of Ni and Mo to obtain a metallic content of 14 wt% of MoO_3 and 3.1 wt% of NiO. For the impregnation with a solution at pH = 7, ammonium heptamolybdate tetrahydrate (Sigma-Aldrich, >99.98%, St Louis, MO, USA) and nickel nitrate hexahydrate (J.T. Baker, >99%, Phillipsburg, NJ, USA) were dissolved in deionized water. For the impregnation with a solution at pH = 9, a solution of NH_4OH was used to adjust the pH. The impregnated samples were thermally treated at 373 K for 240 min, followed by treatment at 823 K (5 K min^{-1}) for 240 min. Catalysts were activated by sulfidation at 673 K for 240 min in a continuous reactor and a gas flow of 3.5×10^{-2} L/min (P_{H_2} = 684 mmHg and P_{H_2S} = 76 mmHg). The catalysts were labeled NiMo/SG7 and NiMo/SG9 for the samples impregnated at pH = 7 or 9, respectively. The endings -D, -C, or -S were employed to denote dried, calcined, or sulfided samples, respectively.

2.3. Characterization Techniques

Samples were analyzed through zeta potential technique (Malvern Zeta90 instrument, Malvern Instruments, Worcestershire, UK). For this purpose, 10 mg of catalyst was suspended in 0.1 L of an aqueous electrolytic solution (0.01 mol L^{-1} $NaNO_3$) and treated with ultrasound for 20 min. For N_2 physisorption analysis (ASAP 2020 Micromeritics instrument, Norcross, GA, USA), samples were thermally pretreated at 573 K for 240 min under vacuum at P = 3×10^{-5} mm Hg. Additionally, the solids were characterized by UV-vis diffuse reflectance spectroscopy using an integration sphere coupled to a spectrophotometer (Lambda 35 PerkinElmer, Waltham, MA, USA), with data presented using the Kubelka–Munk expression as reported previously by [25]. FT-IR (PerkinElmer frontier apparatus with ATR, Waltham, MA, USA) and Raman (BWTEK iRamanPlus spectrometer, B&W TEK, Newark, DE, USA, equipped with an HQE-CCD detector, 532 nm laser, and microscope) spectroscopies were also conducted. Sulfided catalysts were characterized through high resolution electronic microscopy (Thermo Fisher Tecani G2 microscope, Thermo Fisher Scientific, Eindhoven, The Netherlands, 300 kV). For statistical analysis, 10–12 micrographs (2.356×10^{-15} m^2 each) from several regions of the samples

(700 counted particles) were analyzed. The average stacking number was obtained by Equation (1) and the average slab length was obtained by Equation (2) as follows:

$$\overline{N} = \frac{\sum_{i=1}^{n} n_i S_i}{\sum_{i=1}^{n} n_i} \tag{1}$$

$$\overline{L} = \frac{\sum_{i=1}^{n} n_i l_i}{\sum_{i=1}^{n} n_i} \tag{2}$$

where S_i represents the stacking number, l_i is the length of the MoS_2 slab (both determined from micrographs), and n represents the number of particles measured within a stacking number of index i.

2.4. Catalytic Evaluation

HVO tests were performed in a microreactor of the fixed-bed type. For this, 0.1 g of the sulfided sample was packed between two beds of inert Al_2O_3 (Sigma Aldrich, 99%, St Louis, MO, USA). A liquid stream of 6.7×10^{-6} L s^{-1} (10 wt% soybean oil and 0.05 wt% dimethyl disulfide (Sigma-Aldrich, >99%, St Louis, MO, USA) dissolved in n-heptane (Sigma-Aldrich, 99%) and a hydrogen (INFRA, 99.9%) gas stream of 5.8×10^{-4} L s^{-1} were mixed and fed to the reactor. Catalytic evaluation was conducted at 40 bar H_2 and 390 °C for 720 min to achieve stable activity. Reactions aliquots were analyzed via FT-IR (PerkinElmer, Frontier, Waltham, MA, USA) and gas chromatography (PerkinElmer, Autosystem XL, Shelton, CT, USA). The quantification of triglyceride and fatty acid concentrations was calculated following the procedure reported by Rivera-Guasco [17].

The hydrogenolysis constant rate (k_1) was calculated using Equations (3) and (4) as follows:

$$x_1 = \frac{Tg_0 - Tg_x}{Tg_0} \tag{3}$$

$$k_1 = -\frac{F_0}{m_c Tg_0} ln(1 - x_1) \tag{4}$$

where x_1 represents the triglycerides conversion, Tg_0 the initial molar concentration (mol L^{-1}) of triglycerides, Tg_x the molar concentration (mol L^{-1}) of triglycerides, F_0 the initial molar flow (mol s^{-1}) of triglycerides, and m_c the catalyst mass (g).

The deoxygenation constant (k_2) was determined using Equations (5) and (6) as follows:

$$x_2 = \frac{FA_0 - FA_X}{FA_0} \tag{5}$$

$$k_2 = -\frac{F_{F_0}}{m_c FA_0} ln(1 - x_2) \tag{6}$$

where x_2 represents the fatty acids conversion, FA_0 the initial molar concentration (mol L^{-1}) of fatty acids, FA_X the molar concentration (mol L^{-1}) of fatty acids, and F_{F_0} the initial molar flow (mol s^{-1}) of fatty acids.

A commercial $NiMo/Al_2O_3$ catalyst served as the reference for catalytic activity: MoO_3 content = 14 wt% NiO content = 3 wt%, Vp = 0.48 cm^3 g^{-1}, and specific surface area = 250 m^2 g^{-1}.

In the green diesel obtained, the selectivity of hydrocarbons (S_i) was calculated as follows:

$$S_i = \frac{A_i}{\sum A_i} \tag{7}$$

where A_i is the area of the peak corresponding to the obtained i alkane (n-C_{18}, n-C_{17}, n-C_{16}, or n-C_{15}).

3. Results

3.1. Surface Characterization

The Al_2O_3 support was characterized by using zeta potential and N_2 physisorption, and the results are shown in Figure 2. The amphoteric character of the Al_2O_3 surface was analyzed by the zeta potential, Figure 2a. A zeta potential value of zero indicates equilibrium between the solution charges and the surface charge of the support [32]. At pH values between 2 and 7, the positive zeta potential values observed indicate that the Al_2O_3 surface possesses a positive charge ($Al\text{-}OH^{2+}$) in the presence of an acidic solution (H^+). At pH values between 8 and 12, the negative zeta potential values indicate that the surface becomes negatively charged ($Al(OH)^{2-}$ or $AlOO^-$) due to the presence of OH^- ions from the solution [31,32]. The net surface pH of the Al_2O_3 support is 7.8, as shown in Figure 2a. This value is comparable to the one reported by [33]. Based on this, it is suggested that in an impregnation solution at pH = 7, the Al_2O_3 surface is positively charged, allowing interaction with the negative ions in the solution. In contrast, in an impregnation solution at pH = 9, the surface is negatively charged, favoring interaction with the metallic cations in the solution.

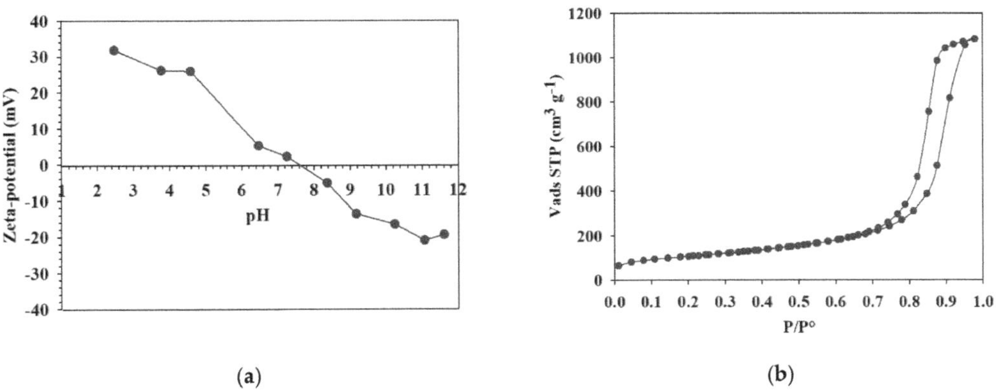

Figure 2. Al_2O_3 surface characterization: (**a**) zeta potential curve; (**b**) N_2 physisorption isotherm.

Figure 2b shows the N_2 physisorption results of the Al_2O_3 sol–gel. The isotherm presents a type IV profile and an H1 hysteresis loop, usually associated with mesoporous solids composed of cylindric channels or agglomerates of spheroidal particles [34]. From the adsorption isotherm, the following textural properties were obtained: porous diameter: 18 nm, porous volume: 1.7 cm^3/g, and specific surface area: 377 m^2/g.

3.2. Catalyst Characterization at the Oxide State

3.2.1. UV-Vis DRS Analysis

Alumina was impregnated with two solutions at pH = 7 and 9 to examine the interactions between Ni, Mo, and the alumina surface. The solution at pH = 7 mainly contains a mixture of heptamolybdate ions ($Mo_7O_{24}{}^{6-}$) and Ni complexes with octahedral local symmetry [$Ni^{2+}6O^{2-}$], located between 650 and 756 nm, and commonly referred to as $Ni^{2+}{}_{(Oh)}$. On the other hand, in the solution at pH = 9, the predominant species are $MoO_4{}^{2-}$

and nickel with tetrahedral local symmetry [$Ni^{2+}4O^{2-}$], located at 623 nm, referred to as $Ni^{2+}{}_{(Td)}$ [27,31].

Figure 3 shows the UV-vis DR spectra of the catalysts after the drying and calcination steps. The alumina support shows no bands. A broad band with a maximum at 300 nm is observed in the NiMo/SG7-D catalyst, corresponding to the $O^{2-} \rightarrow Mo^{6+}$ charge transfer [35], Figure 3a. The band at 370 nm corresponds to the surface species [$Ni^{2+}6H_2O$] [36]. The 650 nm band is related to the $Ni^{2+}{}_{(Oh)}$ surface species. According to the zeta potential results at pH = 7, the alumina possesses a positive surface charge (+2 mV), which is sufficient for the deposition of ions.

After calcination, the NiMo/SG7-C solid exhibits bands at 320 and 360 nm, corresponding to the $O^{2-} \rightarrow Mo^{6+}$ charge transfer, while the band at 390 nm can be associated with the Ni^{2+}/Al_2O_3 species formed from the calcination of [$Ni^{2+}6H_2O$]/Al_2O_3 [35–37], as shown in Figure 3a. The formation of nickel with octahedral distorted symmetry, $Ni^{2+}{}_{(oh\text{-}dis)}$, is indicated by the presence of bands at 720 and 820 nm. These bands can be associated with the Ni-Mo interaction [31,32]. For the impregnation at pH = 9, alumina shows a zeta potential value of −10 mV (Figure 2), indicating that the surface is negatively charged with $Al(OH)^{2-}$ or $AlOO^-$ species.

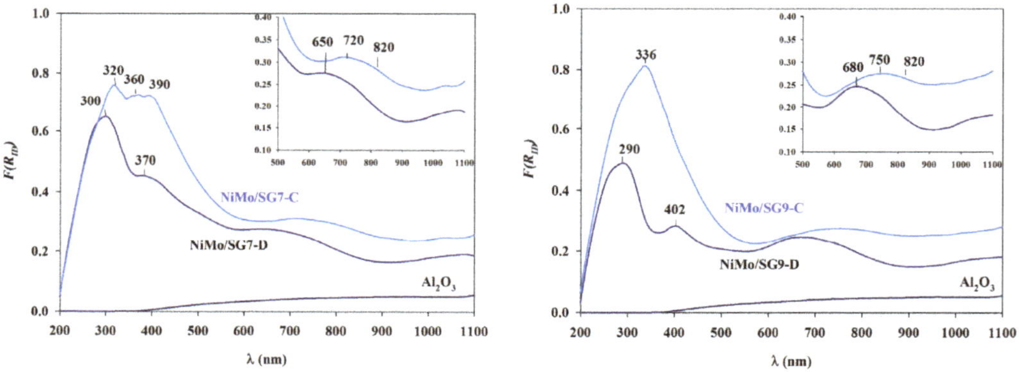

Figure 3. UV-vis diffuse reflectance spectra of (**a**) NiMo/SG7; (**b**) NiMo/SG9 catalysts, dried and calcined samples.

The RD-UV-vis spectrum of the NiMo/SG9-D catalyst shows a band at 290 nm due to the $O^{2-} \rightarrow Mo^{6+}$ charge transfer (Figure 3b). The band at 402 nm corresponds to the Ni^{2+}/Al_2O_3 spinel, which is favored by the presence of $Al(OH)^{2-}$ or $AlOO^-$ surface species. The band at 680 nm is associated with $Ni^{2+}{}_{(Oh)}$. The calcined catalyst NiMo/SG9-C shows an intense band with a maximum of 336 nm, as shown in Figure 3b. This band can be attributed to both the $O^{2-} \rightarrow Mo^{6+}$ charge transfer and the Ni^{2+}/Al_2O_3 spinel. The bands at 750 and 820 nm suggest that the $Ni^{2+}{}_{(oh\text{-}dis)}$ interacts with Mo species [38,39].

3.2.2. FT-IR Analysis

In the supported metal oxide catalysts, spectroscopy analysis provides complementary information regarding the surface species formed [40]. FT-IR spectroscopy allows the analysis of the vibrational modes associated with chemical interactions within the catalyst structures [41], particularly the vibrations of the terminal Mo=O and bridging Mo-O-Mo. The FT-IR spectra of the NiMo/SG7 and NiMo/SG9 catalysts after the drying and calcination steps are shown in Figure 4. All catalysts show a band at 540 cm^{-1}, associated with the vibrations of Mo-O-Mo bridges [42]. Additionally, all samples exhibit a wide band between 820 and 1000 cm^{-1}, which is associated with the Mo=O and Mo-O vibrations of surface metal oxide species [40]. Both dried samples, NiMo/SG7-D and NiMo/SG-9-D, exhibit two broad bands at 1320 and 1410 cm^{-1}, corresponding to ionic nitrates from the precursors [43]. After calcination, these bands are no longer detected in the NiMo/SG7-C and NiMo/SG9-C samples.

3.2.3. Raman Analysis

Catalysts were also characterized by Raman spectroscopy, as shown in Figure 5. The NiMo/SG7-D catalyst shows a band at 942 cm^{-1}, corresponding to the Mo=O_t vibration of the $Mo_7O_{24}^{6-}$ species [27,31,44], as shown in Figure 5a. This suggests that the positive charge of Al_2O_3 during the impregnation at pH = 7 facilitates the adsorption of these species. The NiMo/SG9-D catalyst shows a band at 935 cm^{-1}, attributed to the Mo=O_t vibration of the $Mo_7O_{24}^{6-}$ species, as shown in Figure 5a. It is important to note that the impregnation solution at pH = 9 mainly contains MoO_4^{2-} species, which polymerize into $Mo_7O_{24}^{6-}$ species upon deposition onto the support.

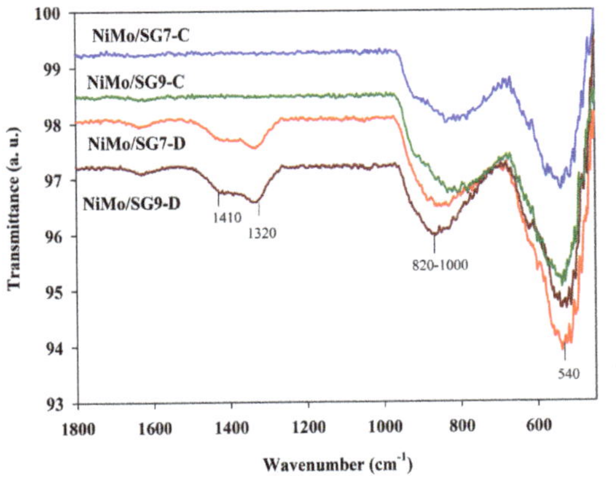

Figure 4. FT-IR spectra of NiMo/SG7 and NiMo/SG9 catalysts, dried and calcined samples.

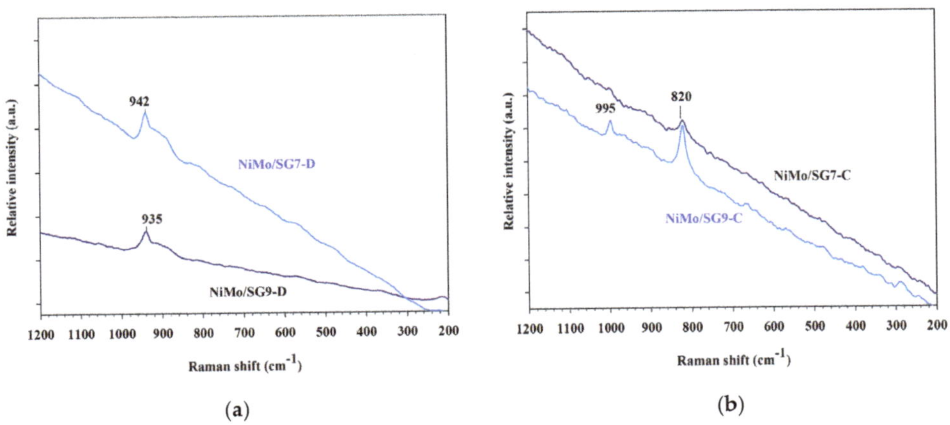

Figure 5. Raman spectra of (**a**) NiMo/SG7; (**b**) NiMo/SG9 catalysts, dried and calcined samples.

After calcination, the NiMo/SG9-C sample exhibits two bands at 820 and 995 cm^{-1}, which are associated with Mo-O-Mo and Mo=O_t vibrations of the orthorhombic MoO_3 cluster [44,45], as shown in Figure 5b. This result suggests that during impregnation at pH = 9, the formation of the molybdenum cluster is favored by the negative charge on the alumina surface. The NiMo/SG7-C catalyst shows the band at 820 cm^{-1}, which is associated with an amorphous molybdenum trioxide cluster [46], as shown in Figure 5b. In this sample, the band at 995 cm^{-1} has low intensity due to the formation of a Ni-O-Mo interaction.

Based on the UV-vis, FT-IR, and Raman results, impregnation at pH = 9 induces a negative charge on the Al_2O_3 surface, resulting in the formation of Ni^{2+}/Al_2O_3, $Ni^{2+}{}_{(Oh\text{-}dis)}$, and MoO_3 surface species. In contrast, impregnation at pH = 7 results in a positive charge of the Al_2O_3 surface, leading to the formation of surface species such as $Ni^{2+}{}_{(Oh)}$ and $Ni^{2+}{}_{(oh\text{-}dis)}$, which interact with non-crystalline molybdenum trioxide.

3.3. *Catalyst Characterization at the Sulfide State*

TEM Analysis

The influence of the impregnation pH on the morphology of MoS_2 particles was analyzed by TEM. Figure 6 shows the micrographs of the NiMo/SG7-S and NiMo/SG9-S catalysts. The dispersion of MoS_2 particles on both catalysts is relatively homogeneous, consisting of one to two slabs and showing the characteristic fringes of the MoS_2 phase. The number of stacked particles per nm^2, stacking degrees, and average slab lengths are shown in Table 1.

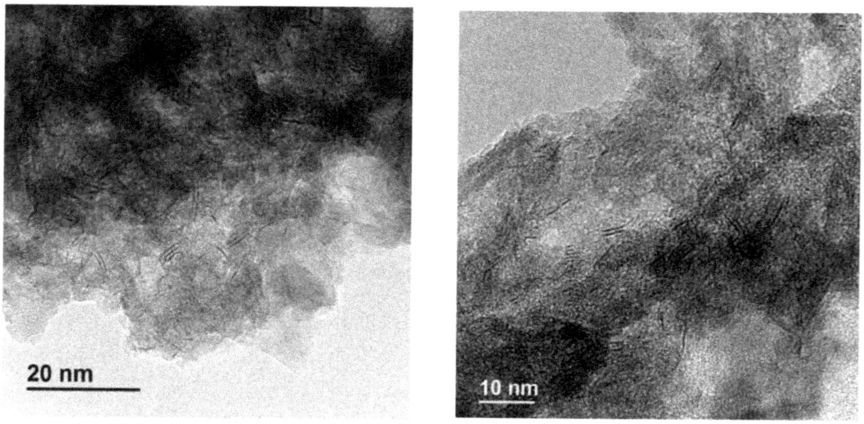

(a) (b)

Figure 6. TEM representative micrographs of (**a**) NiMo/SG7-S; (**b**) NiMo/SG9-S catalysts, sulfided samples.

Table 1. Density of stacks per 1000 nm^2, average stacking number (N), and average stack length (L) of catalysts.

Catalyst	Stacks per 1000 nm^2	N	L (nm)
NiMo/SG7-S	10.9	1.56	2.91
NiMo/SG9-S	10.0	1.54	2.79

According to TEM analysis, the number of stacks per 1000 nm^2 is very similar between both samples, with values of 10.9 and 10.0 for the NiMo/SG7 and NiMo/SG9 catalysts, respectively. This result suggests that the impregnation pH does not significantly affect the stacking density in these catalysts. The stacking degree of MoS_2 particles is also not influenced by the impregnation pH, remaining within a narrow range: 1.56 for NiMo/SG7-S and 1.54 for NiMo/SG9-S. The most significant change observed is in the average slab, where impregnation at pH = 7 favors the formation of longer slabs (2.91 nm) compared to those formed at pH = 9 (2.79 nm).

3.4. Hydroprocessing of Soybean Oil

The impact of the impregnation pH on the activity of sulfided catalysts was analyzed in the hydroprocessing of soybean oil. The results for the hydrogenolysis (k_1) and deoxygenation (k_2) rate constants are summarized in Figure 7. In the first step, the NiMo/SG7-S catalyst exhibits the higher hydrogenolysis rate constant (k_1), with a value of 48.2×10^{-5} $Ls^{-1}g^{-1}$, as shown in Figure 7a. This catalyst exhibits a K_1 value 13% higher than the reference catalyst (42.6 $Ls^{-1}g^{-1}$). One can notice that the NiMo/SG7-S catalyst shows a k_1 value 17% higher than the NiMo/SG9-S catalyst (41.1×10^{-5} $Ls^{-1}g^{-1}$). This result suggests that the presence of the Ni^{2+}/Al_2O_3 species reduces the activity of the NiMo/SG9-S solid. On the other hand, deoxygenation rate constants (k_2) exhibit values higher than those obtained in the hydrogenolysis step, as shown in Figure 7b. These results suggest that green diesel production is mainly controlled by the hydrogenolysis mechanism. Figure 7b shows that the NiMo/SG7-S catalyst exhibits the highest deoxygenation rate constant (k_2), with a value of 16.3×10^{-4} $Ls^{-1}g^{-1}$. This value is even higher than the one observed for the NiMo/SG9-S catalyst (12.8×10^{-4} $Ls^{-1}g^{-1}$). In the deoxygenation step, the NiMo/SG7-S catalyst shows a k_2 value higher than the one observed for the reference catalyst (11.8×10^{-4} $Ls^{-1}g^{-1}$), suggesting that the NiMo/SG7-S sample has the potential to serve as a catalyst in green diesel production.

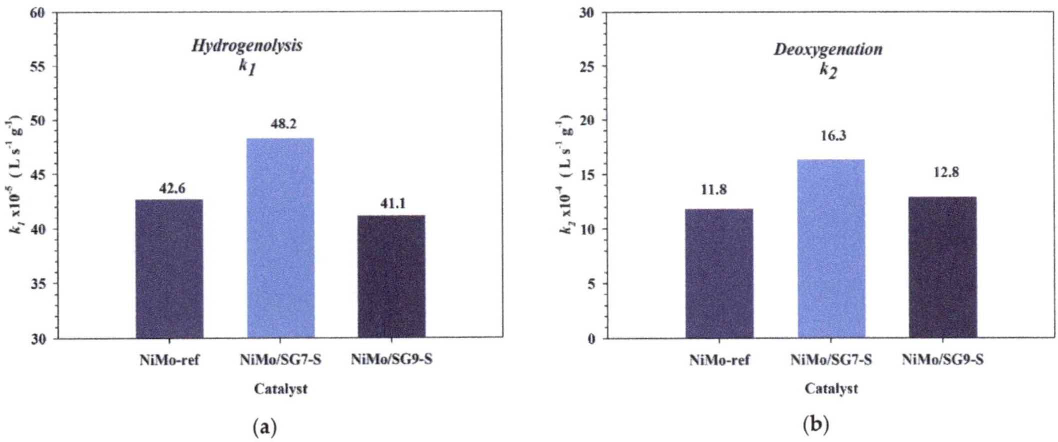

Figure 7. Hydroprocessing of soybean oil: (**a**) hydrogenolysis (k_1) rare constants and (**b**) deoxygenation (k_2) rate constants for the different sulfided catalysts at 390 °C.

The green diesel yield of each catalyst is shown in Figure 8. The NiMo/SG9-S catalyst exhibits a similar yield (27.1%) to the reference catalyst (25.4%). However, the NiMo/SG7-S catalyst shows the highest yield value (37.1%). This result is consistent with the rate values observed previously.

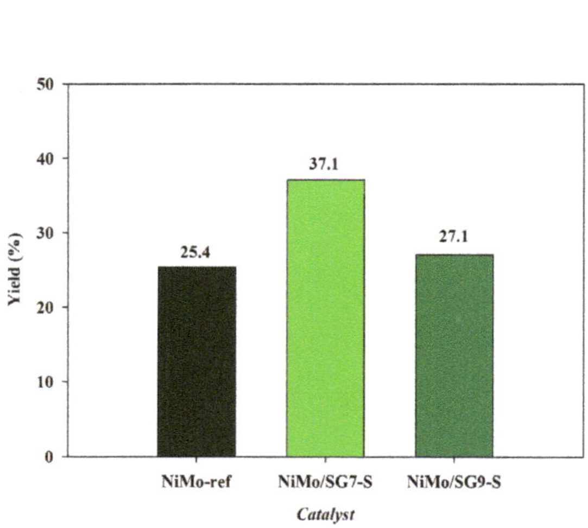

Figure 8. Green diesel yield for the NiMo-reference, NiMo/SG7-S, and NiMo/SG9-S catalysts.

Reaction products were analyzed by gas chromatography. The selectivity results at 390 °C are shown in Table 2. All samples exhibit very high selectivity to n-C_{17}, with values of 77.3, 75.5, and 74.3 mol% for NiMo/Al_2O_3-S, NiMo/SG7-S, and NiMo/SG9-S samples, respectively. Notably, both NiMo/SG7-S and NiMo/SG-9-S catalysts show higher selectivity to n-C_{18} (8.3 and 10.3% mol, respectively) compared to the reference sample (7.3%mol). In HVO, n-C_{18} is considered the most valuable product [47]. Furthermore, the NiMo/SG7-S sample exhibits higher selectivity to n-C_{15} (10.6%mol) than the NiMo/SG9-S catalyst (9.2%mol).

Table 2. Hydrocarbon selectivity at 390 °C.

Catalyst	n-C_{18} (mol%)	n-C_{17} (mol%)	n-C_{16} (mol%)	n-C_{15} (mol%)
NiMo/Al_2O_3-S	7.3	77.3	4.8	10.3
NiMo/SG7-S	8.3	75.5	5.3	10.6
NiMo/SG9-S	10.3	74.3	5.9	9.2

4. Discussion

4.1. Effect of NiMo Surface Species on the Hydroprocessing of Soybean Oil

In this work, the Ni and Mo surface species generated during the catalyst synthesis were analyzed. Impregnations with solutions at pH = 7 and 9 allowed for the identification of the Ni-Mo-Al_2O_3 interactions. These interactions are crucial for understanding the behavior of the catalysts during hydroprocessing reactions. The results show that impregnation at pH = 7 causes the Al_2O_3 sol–gel to become positively charged, leading to the formation of the $Ni^{2+}{}_{(Oh)}/Mo_7O_{24}{}^{6-}$ surface species. These species are transformed into $Ni^{2+}{}_{(Oh)}$/molybdenum trioxide (non-crystalline), after the calcination step. In contrast, impregnation at pH = 9 results in a negatively charged alumina surface, leading to the formation of $Ni^{2+}{}_{(Oh)}$ and $Mo_7O_{24}{}^{6-}$. However, after calcination, the formation of $Ni^{2+}{}_{(Oh)}$-crystalline molybdenum trioxide and the Ni^{2+}/Al_2O_3 species becomes evident.

In the sulfide state, the activity of the catalysts is influenced by the formation of the so-called NiMoS active phases [9,48], where Ni interacts with the MoS_2 phase. According to HRTEM results, similar dispersion and length of the MoS_2 slabs were observed for both catalysts. However, a deeper analysis of the spectroscopy results suggests that the

NiMo/SG9-S catalyst exhibits the formation of the Ni^{2+}/Al_2O_3 spinel after the drying and calcination steps. These species limit the availability of Ni to form the NiMoS active phases because the nickel interacts with the support instead of with Mo. Consequently, a decrease in HVO activity is observed, as shown in Figure 7a.

Finally, the green diesel produced through the combination of sol–gel alumina and impregnation at pH = 7 results in the development of a catalyst that is more active than the reference catalyst.

4.2. Future Research Directions

The hydroprocessing of vegetable oil (HVO) is performed in a hydroprocessing unit, which involves multiple variables. As a result, understanding and optimizing all the parameters involved are continuous challenges. Future research should focus on several aspects to improve efficiency:

1. Catalyst design and optimization: A crucial aspect of improving HVO production is the development of more efficient catalysts. This includes investigating metal support interactions, optimizing metal loading, and exploring the use of other types of promotors or alternative active phases.
2. Alternative feedstocks: The use of alternative feedstocks, such as canola or palm oils, may improve the quality of green diesel. Moreover, using lignocellulosic biomass for biofuel production offers the potential for increasing sustainability.
3. Waste feedstocks: The use of waste feedstocks, such as used cooking oils, could provide a more cost-effective alternative for HVO.

5. Conclusions

The impact of impregnation pH on the surface species of Ni and Mo was studied during both the catalyst synthesis and the hydroprocessing of soybean oil. Although different types of catalysts have been used for biofuel production, this study emphasizes that a deeper understanding of the surface species formed provides the basis for proposing and designing strategies for catalyst optimization, favoring the efficient production of green diesel.

Author Contributions: Conceptualization, A.G.-L.; methodology, A.G.M.-A.; software, A.G.M.-A.; validation, J.A.R.-Á.; formal analysis, A.G.-L. and A.L.-B.; investigation, A.G.M.-A.; resources, A.G.-L.; data curation, A.G.-L.; writing—original draft preparation, A.G.-L. and A.L.-B.; writing—review and editing, A.G.-L. and A.L.-B.; visualization, R.S.-R.; supervision, J.A.R.-Á.; project administration, A.G.-L.; funding acquisition, A.G.-L. All authors have read and agreed to the published version of the manuscript.

Funding: This research received no external funding.

Institutional Review Board Statement: Not applicable.

Informed Consent Statement: Not applicable.

Data Availability Statement: The data are contained within the article.

Acknowledgments: Alexis G. Mijares-Alvarado is very grateful to SECIHTI through the "Convocatoria de Becas Nacionales para Estudios de Posgrado" for the doctoral grant. Acela López-Benítez thanks SECIHTI for the postdoctoral grant.

Conflicts of Interest: The authors declare no conflicts of interest.

References

1. Lucantonio, S.; Di Giuliano, A.; Rossi, L.; Gallucci, K. Green Diesel Production via Deoxygenation Process: A Review. *Energies* **2023**, *16*, 844. [CrossRef]
2. Chia, S.R.; Nomanbhay, S.; Chew, K.W.; Show, P.L.; Milano, J.; Shamsuddin, A.H. Indigenous Materials as Catalyst Supports for Renewable Diesel Production in Malaysia. *Energies* **2022**, *15*, 2835. [CrossRef]
3. Khan, S.; Paliwal, V.; Pandey, V.V.; Kumar, V. Biomass as Renewable Energy. *Int. Adv. Res. J. Sci. Eng. Technol.* **2015**, *2*, 301–304.
4. Perlack, R.D.; Wright, L.L.; Turhollow, A.F.; Graham, R.L.; Stoke, B.J.; Erbach, D.C. *Biomass as a Feedstock for a Bioenergy and Bioproducts Industry: The Technical Feasibility of a Billion-Ton Annual Supply*; Oak Ridge National Laboratory: Oak Ridge, TN, USA, 2005. [CrossRef]
5. Hongloi, N.; Prapainainar, P.; Prapainainar, C. Review of Green Diesel Production from Fatty Acid Deoxygenation over Ni-Based Catalysts. *Mol. Catal.* **2022**, *523*, 111696. [CrossRef]
6. Mahdi, H.I.; Bazargan, A.; McKay, G.; Azelee, N.I.W.; Meili, L. Catalytic Deoxygenation of Palm Oil and Its Residue in Green Diesel Production: A Current Technological Review. *Chem. Eng. Res. Des.* **2021**, *174*, 158–187. [CrossRef]
7. Li, D.-C.; Pan, Z.; Tian, Z.; Zhang, Q.; Deng, X.; Jiang, H.; Wang, G.-H. Frustrated Lewis Pair Catalyst Realizes Efficient Green Diesel Production. *Nat. Commun.* **2024**, *15*, 3172. [CrossRef]
8. Knothe, G. Biodiesel and Renewable Diesel: A Comparison. *Prog. Energy Combust. Sci.* **2010**, *36*, 364–373. [CrossRef]
9. Wagenhofer, M.F.; Baráth, E.; Gutiérrez, O.Y.; Lercher, J.A. Carbon-Carbon Bond Scission Pathways in the Deoxygenation of Fatty Acids on Transition-Metal Sulfides. *ACS Catal.* **2017**, *7*, 1068–1076. [CrossRef]
10. Jęczmionek, Ł.; Krasodomski, W. Hydroconversion of Vegetable Oils Isomerized over ZSM-5: Composition and Properties of Hydroraffinates. *Energy Fuels* **2015**, *29*, 3739–3747. [CrossRef]
11. Durgut, M.R. Optimized Biodiesel Production from Pumpkin (*Cucurbita pepo* L.) Seed Oil: A Response Surface Methodology for Microwave-Assisted Transesterification. *Processes* **2025**, *13*, 572. [CrossRef]
12. Huber, G.W.; O'Connor, P.; Corma, A. Processing Biomass in Conventional Oil Refineries: Production of High Quality Diesel by Hydrotreating Vegetable Oils in Heavy Vacuum Oil Mixtures. *Appl. Catal. A Gen.* **2007**, *329*, 120–129. [CrossRef]
13. Kim, S.K.; Han, J.Y.; Lee, H.S.; Yum, T.; Kim, Y.; Kim, J. Production of Renewable Diesel via Catalytic Deoxygenation of Natural Triglycerides: Comprehensive Understanding of Reaction Intermediates and Hydrocarbons. *Appl. Energy* **2014**, *116*, 199–205. [CrossRef]
14. Szarvas, T.; Eller, Z.; Kasza, T.; Ollár, T.; Tétényi, P.; Hancsók, J. Radioisotopic Investigation of the Oleic Acid-1-14C HDO Reaction Pathways on Sulfided Mo/P/Al_2O_3 and NiW/Al_2O_3 Catalysts. *Appl. Catal. B* **2015**, *165*, 245–252. [CrossRef]
15. Ameen, M.; Azizan, M.T.; Yusup, S.; Ramli, A.; Yasir, M. Catalytic Hydrodeoxygenation of Triglycerides: An Approach to Clean Diesel Fuel Production. *Renew. Sustain. Energy Rev.* **2017**, *80*, 1072–1088. [CrossRef]
16. Ishihara, A.; Fukui, N.; Nasu, H.; Hashimoto, T. Hydrocracking of Soybean Oil Using Zeolite–Alumina Composite Supported NiMo Catalysts. *Fuel* **2014**, *134*, 611–617. [CrossRef]
17. Rivera-Guasco, R.; López-Benítez, A.; Guevara-Lara, A. Ni-W Catalysts Supported on TiO_2-Al_2O_3 for Efficient Green Diesel Production: "Standard" vs "Keggin" Comparison. *ACS Omega* **2025**, *10*, 4276–4290. [CrossRef] [PubMed]
18. Itthibenchapong, V.; Srifa, A.; Kaewmeesri, R.; Kidkhunthod, P.; Faungnawakij, K. Deoxygenation of Palm Kernel Oil to Jet Fuel-like Hydrocarbons Using Ni-MoS_2/γ-Al_2O_3 Catalysts. *Energy Convers. Manag.* **2017**, *134*, 188–196. [CrossRef]
19. Liu, S.; Zhu, Q.; Guan, Q.; He, L.; Li, W. Bio-Aviation Fuel Production from Hydroprocessing Castor Oil Promoted by the Nickel-Based Bifunctional Catalysts. *Bioresour. Technol.* **2015**, *183*, 93–100. [CrossRef]
20. Hancsók, J.; Kasza, T.; Kovács, S.; Solymosi, P.; Holló, A. Production of Bioparaffins by the Catalytic Hydrogenation of Natural Triglycerides. *J. Clean. Prod.* **2012**, *34*, 76–81. [CrossRef]
21. Kumar, R.; Farooqui, S.A.; Anand, M.; Kumar, R.; Joshi, R.; Khan, A.; Sinha, A.K. Hydrotreatment of Jatropha Oil over NiMoS Catalyst Supported on Thermostable Mesoporous Silica Doped Titania for the Production of Renewable Drop-in Diesel. *Catal. Commun.* **2017**, *98*, 102–106. [CrossRef]
22. Kordouli, E.; Sygellou, L.; Kordulis, C.; Bourikas, K.; Lycourghiotis, A. Probing the Synergistic Ratio of the NiMo/γ-Al_2O_3 Reduced Catalysts for the Transformation of Natural Triglycerides into Green Diesel. *Appl. Catal. B* **2017**, *209*, 12–22. [CrossRef]
23. Liu, Y.; Sotelo-Boyás, R.; Murata, K.; Minowa, T.; Sakanishi, K. Hydrotreatment of Vegetable Oils to Produce Bio-Hydrogenated Diesel and Liquefied Petroleum Gas Fuel over Catalysts Containing Sulfided Ni–Mo and Solid Acids. *Energy Fuels* **2011**, *25*, 4675–4685. [CrossRef]
24. Mikulec, J.; Cvengroš, J.; Joríková, Ľ.; Banič, M.; Kleinová, A. Second Generation Diesel Fuel from Renewable Sources. *J. Clean. Prod.* **2010**, *18*, 917–926. [CrossRef]
25. López-Benítez, A.; Berhault, G.; Guevara-Lara, A. NiMo Catalysts Supported on Mn-Al_2O_3 for Dibenzothiophene Hydrodesulfurization Application. *Appl. Catal. B* **2017**, *213*, 28–41. [CrossRef]

26. Bui, N.Q.; Geantet, C.; Berhault, G. Maleic Acid, an Efficient Additive for the Activation of Regenerated CoMo/Al_2O_3 Hydrotreating Catalysts. *J. Catal.* **2015**, *330*, 374–386. [CrossRef]
27. López-Benítez, A.; Berhault, G.; Guevara-Lara, A. Addition of Manganese to Alumina and Its Influence on the Formation of Supported NiMo Catalysts for Dibenzothiophene Hydrodesulfurization Application. *J. Catal.* **2016**, *344*, 59–76. [CrossRef]
28. Fan, Y.; Xiao, H.; Shi, G.; Liu, H.; Qian, Y.; Wang, T.; Gong, G.; Bao, X. Citric Acid-Assisted Hydrothermal Method for Preparing NiW/USY–Al_2O_3 Ultradeep Hydrodesulfurization Catalysts. *J. Catal.* **2011**, *279*, 27–35. [CrossRef]
29. Eijsbouts, S.; Van Den Oetelaar, L.C.A.; Van Puijenbroek, R.R. MoS2 Morphology and Promoter Segregation in Commercial Type 2 Ni–Mo/Al_2O_3 and Co–Mo/Al_2O_3 Hydroprocessing Catalysts. *J. Catal.* **2005**, *229*, 352–364. [CrossRef]
30. López-Benítez, A.; Guevara-Lara, A.; Berhault, G. Nickel-Containing Polyoxotungstates Based on $[PW_9O_{34}]^{9-}$ and $[PW_{10}O_{39}]^{13-}$ Keggin Lacunary Anions Supported on Al_2O_3 for Dibenzothiophene Hydrodesulfurization Application. *ACS Catal.* **2019**, *9*, 6711–6727. [CrossRef]
31. Guevara-Lara, A.; Bacaud, R.; Vrinat, M. Highly Active NiMo/TiO_2-Al_2O_3 Catalysts: Influence of the Preparation and the Activation Conditions on the Catalytic Activity. *Appl. Catal. A Gen.* **2007**, *328*, 99–108. [CrossRef]
32. Hunter, R.J. Chapter 2—Charge and Potential Distribution at Interfaces. In *Zeta Potential in Colloid Science*; Hunter, R.J., Ed.; Academic Press: Cambridge, MA, USA, 1981; pp. 11–58, ISBN 978-0-12-361961-7.
33. Khaleel, A.; Al-Mansouri, S. Meso-Macroporous γ-Alumina by Template-Free Sol–Gel Synthesis: The Effect of the Solvent and Acid Catalyst on the Microstructure and Textural Properties. *Colloids Surf. A Physicochem. Eng. Asp.* **2010**, *369*, 272–280. [CrossRef]
34. Thommes, M.; Kaneko, K.; Neimark, A.V.; Olivier, J.P.; Rodriguez-Reinoso, F.; Rouquerol, J.; Sing, K.S.W. Physisorption of Gases, with Special Reference to the Evaluation of Surface Area and Pore Size Distribution (IUPAC Technical Report). *Pure Appl. Chem.* **2015**, *87*, 1051–1069. [CrossRef]
35. Fournier, M.; Thouvenot, R.; Rocchiccioli-Deltcheff, C. Catalysis by Polyoxometalates. Part 1—Supported Polyoxoanions of the Keggin Structure: Spectroscopic Study (IR, Raman, UV) of Solutions Used for Impregnation. *J. Chem. Soc. Faraday Trans.* **1991**, *87*, 349–356. [CrossRef]
36. Iova, F.; Trutia, A. On the Structure of the NiO–Al_2O_3 Systems, Studied by Diffuse-Reflectance Spectroscopy. *Opt. Mater.* **2000**, *13*, 455–458. [CrossRef]
37. Jocono, M.L.; Schiavello, M.; Cimino, A. Structural, Magnetic, and Optical Properties of Nickel Oxide Supported on η- and η-γ-Aluminas. *J. Phys. Chem.* **1971**, *75*, 1044–1050. [CrossRef]
38. Lepetit, C.; Che, M. Discussion on the Coordination of Ni^{2+} Ions to Lattice Oxygens in Calcined Faujasite-Type Zeolites Followed by Diffuse Reflectance Spectroscopy. *J. Phys. Chem.* **1996**, *100*, 3137–3143. [CrossRef]
39. Teixeira da Silva, V.L.S.; Frety, R.; Schmal, M. Activation and Regeneration of a NiMo/Al_2O_3 Hydrotreatment Catalyst. *Ind. Eng. Chem. Res.* **1994**, *33*, 1692–1699. [CrossRef]
40. Wachs, I.E. Raman and IR Studies of Surface Metal Oxide Species on Oxide Supports: Supported Metal Oxide Catalysts. *Catal. Today* **1996**, *27*, 437–455. [CrossRef]
41. Wachs, I.E. Infrared Spectroscopy of Supported Metal Oxide Catalysts. *Colloids Surf. A Physicochem. Eng. Asp.* **1995**, *105*, 143–149. [CrossRef]
42. Imamura, S.; Sasaki, H.; Shono, M.; Kanai, H. Structure of Molybdenum Supported on α-, γ-, and χ-Aluminas in Relation to Its Epoxidation Activity. *J. Catal.* **1998**, *177*, 72–81. [CrossRef]
43. Nova, I.; Castoldi, L.; Lietti, L.; Tronconi, E.; Forzatti, P.; Prinetto, F.; Ghiotti, G. NOx Adsorption Study over Pt–Ba/Alumina Catalysts: FT-IR and Pulse Experiments. *J. Catal.* **2004**, *222*, 377–388. [CrossRef]
44. Seguin, L.; Figlarz, M.; Cavagnat, R.; Lassègues, J.C. Infrared and Raman Spectra of MoO_3 Molybdenum Trioxides and MoO_3XH_2O Molybdenum Trioxide Hydrates. *Spectrochim. Acta A Mol. Biomol. Spectrosc.* **1995**, *51*, 1323–1344. [CrossRef]
45. Stencel, J.M. *Raman Spectroscopy for Catalysis*; Van Nostrand Reinhold; Springer: New York, NY, USA, 1990.
46. Dieterle, M.; Weinberg, G.; Mestl, G. Raman Spectroscopy of Molybdenum Oxides Part I. Structural Characterization of Oxygen Defects in MoO_{3-x} by DR UV/VIS, Raman Spectroscopy and X-Ray Diffraction. *Phys. Chem. Chem. Phys.* **2002**, *4*, 812–821. [CrossRef]
47. Prangklang, D.; Tumnantong, D.; Yoosuk, B.; Ngamcharussrivichai, C.; Prasassarakich, P. Selective Deoxygenation of Waste Cooking Oil to Diesel-Like Hydrocarbons Using Supported and Unsupported $NiMoS_2$ Catalysts. *ACS Omega* **2023**, *8*, 40921–40933. [CrossRef]
48. Topsøe, H. The Role of Co-Mo-S Type Structures in Hydrotreating Catalysts. *Appl. Catal. A Gen.* **2007**, *322*, 3–8. [CrossRef]

surfaces

Article

Synthesis and Characterization of SnO_2/α-Fe_2O_3, In_2O_3/α-Fe_2O_3, and ZnO/α-Fe_2O_3 Thin Films: Photocatalytic and Antibacterial Applications

Asma Arfaoui [1,*] and Ammar Mhamdi [2,3]

1 Physics Department, College of Science, Jouf University, Sakaka P.O. Box 2014, Saudi Arabia
2 Department of Physics, College of Science and Humanities Dawadmi, Shaqra University, Dawadmi P.O. Box 33, Saudi Arabia; ammar.mhamdi@su.edu.sa
3 Laboratoire de Nanomatériaux, Nanotechnologie et Energie (L2NE), Faculté des Sciences de Tunis, Tunis El Manar University, Tunis P.O. Box 2092, Tunisia
* Correspondence: aarfaoui@ju.edu.sa

Abstract: The fabrication of metal oxide semiconductor heterostructures is a major way to enhance their properties in photocatalytic and antibacterial applications. In the present work, ZnO/α-Fe_2O_3, In_2O_3/α-Fe_2O_3, and SnO_2/α-Fe_2O_3 are chosen to create the heterostructure of thin films using the spray pyrolysis method. This paper compares the experimental results of the structural and morphological properties of the prepared thin layers using XRD, Raman and SEM. The X-ray diffraction shows that the obtained thin film heterostructures crystallize in a hexagonal phase of ZnO, a cubic phase of In_2O_3 and a tetragonal structure of SnO_2, with all of the preceding phases positioned on the rhombohedral phase of the hematite α-Fe_2O_3. In addition, the SEM study provided the morphology and surface structure and confirmed the presence of a highly folded, rough, uneven surface with imperfections of 20 and 65 nm for In_2O_3/α-Fe_2O_3 and SnO_2/α-Fe_2O_3. The photoactivity of the prepared materials was tested via the photocatalytic degradation of methylene blue (MB) dye. Consequently, our findings demonstrate that the cracked surface improves the rapid absorption of contaminants and allows water to easily pass through the surface of the thin layers. Finally, the antibacterial abilities of ZnO/α-Fe_2O_3, In_2O_3/α-Fe_2O_3, and SnO_2/α-Fe_2O_3 thin films were investigated by using the agar well-diffusion technique, comparing the results to the Gram-negative of Pseudomonas aeruginosa and Gram-positive of Bacillus subtilis, and these thin films were found to have high antibacterial activity.

Keywords: biophysics study; chemical spray pyrolysis; structural study; photocatalytic activity; antibacterial test

1. Introduction

Recent years have seen a significant increase in interest in thin film technology because of its many uses, including in gas sensors [1,2], energy storage [3–5], photovoltaic solar cells [6–8], heterogeneous photocatalysis for water splitting processes [9–11], and optoelectronic devices [12–14]. The amount of wastewater polluted with undesirable and dangerous dye has remarkably increased. The colored dyes that are frequently employed in the textile sector are one of the primary causes of pollution. Water's transparency and gas solubility can be impacted by even minute amounts of dye, which can also give the water an incredibly vibrant color [15,16].

As a result, it is now crucial to explore the possibility of functionalizing various surface types without altering their optical characteristics. There exist numerous techniques for obtaining thin films, such as electroplating [17], anodic treatment [18], chemical vapor deposition (CVD) [19,20], atomic layer deposition (ALD) [21,22], and spin coating [23]. The spray pyrolysis process is the most commonly used because it is a simple way to make films with any material in any amount by adding the film to the spray precursor solution.

Owing to its low cost, small bandgap, environmental friendliness, and thermodynamic stability, hematite (Fe_2O_3) is a promising semiconductor material for use as an efficient photocatalyst under visible light irradiation. To increase the photoinduced electron–hole separation and, consequently, the photocatalytic activity of Fe_2O_3, heterojunction construction is used. Because it is an environmentally beneficial technology, the photocatalytic reduction of methylene blue has been applied extensively in a variety of semiconductor materials, including Fe_2O_3, In_2O_3, SnO_2, and ZnO, achieved via exposing them to UV and visible light. In this context, Alofi et al. [24] describe the mechanism of photogenerating radical species through a particulate film, such as $OH^\bullet$ (instead of h^+) and super oxide $O_2^{-\bullet}$ (instead of e^-) radicals. However, it is unclear why these species would not react quickly with one another as they pass through the film from the bulk to the surface. Different kinds of dyes are also investigated as different types of pollution; Singh et al. [25] used doped ZnO as a photocatalyst to decompose methyl orange (MO), methylene blue (MB), and congo red (CR), respectively. The insertion of another material into the oxide semiconductor lattice can alter its optical, morphological, and structural characteristics, among other physical and chemical modifications. These modifications might alter the semiconductor's photocatalytic activities and shift them toward a visible area of the spectrum. Semiconductor heterojunctions, which comprise two or more semiconductors arranged in delayed energy bands, enable the separation of photogenerated electron–holes and the retention of redox capacity. R. M. Mohamed et al. [26] studied the effect of increasing the solar light-driven photocatalytic hydrogen evolution of different semiconductor photocatalysts; this technique might be employed when using organic and non-metal-based new-age materials, especially if combined with numerous modification procedures. The present paper aims to provide a well-founded direct comparison of three thin film heterostructures ($ZnO/\alpha\text{-}Fe_2O_3$, $In_2O_3/\alpha\text{-}Fe_2O_3$, and $SnO_2/\alpha\text{-}Fe_2O_3$) prepared using a simple chemical technique. The photodegradation of methylene bule dye MB and the antibacterial application of the prepared materials were tested and compared. We need to point out that no previous research has been conducted to date using this method to create these thin film heterostructures.

2. Experimental Section

2.1. Film Preparation

2.1.1. ZnO Thin Film Preparation

Zinc acetate ($C_4H_6O_4Zn$, $2H_2O$) was chosen as the primary precursor and was dissolved in isopropyl alcohol at a concentration of 10^{-2} mol/l to create the precursor solution.

Using the chemical spray approach, ZnO thin films were applied to a glass substrate at 460 °C [27].

2.1.2. In_2O_3 Thin Film Preparation

An aqueous solution of indium chloride ($InCl_3$) with a concentration of 0.01 M was sprayed onto glass substrates and heated to 350 °C using a N_2 gas pressure of 0.5 bar to create an In_2O_3 thin layer.

2.1.3. SnO_2 Thin Film Preparation

To create SnO_2 thin films, a precursor was obtained in a 0.05 M solution by dissolving tin chloride pentahydrate ($SnCl_4.5H_2O$) in pure ethanol (C_2H_5OH). This solution was sprayed at 0.5 bar of N_2 gas pressure onto glass substrates that were heated to 450 °C on a hot plate.

2.1.4. ZnO/α-Fe_2O_3, In_2O_3/α-Fe_2O_3, and SnO_2/α-Fe_2O_3 Thin Film Preparation

To prepare ZnO/α-Fe_2O_3, In_2O_3/α-Fe_2O_3, and SnO_2/α-Fe_2O_3 thin film heterostructures, an aqueous solution containing iron (III) and chloride dehydrate ($FeCl_3.2H_2O$) was utilized. After preparing this solution, we sprayed it at 350 °C onto the ZnO, In_2O_3, and SnO_2 substrates for use as a precursor. Using a nozzle with a 0.5 mm diameter, 20 mL of the resultant solution was sprayed at a rate of 4 mL/min. There was a 30 cm distance between the substrate and the spray gun.

2.2. *Characterization Techniques*

Initially, to study the structural properties using Cu Kα radiation (λ = 0.15418 nm) and 2θ varying from 10° to 70°, the X-ray diffraction spectra of the produced thin films were examined using a copper-source diffractometer (Analytical X Pert PROMP D, Sakaka, Jouf). In addition, at room temperature, Raman scattering studies were recorded using the Jobin Yvon Horibra LABRAM-HR micro-Raman system (Sakaka, Jouf), which was detectable within the 200–1200 cm^{-1} range. Subsequently, a scanning electron microscope (SEM) of the JEOL-JSM 5400 type (Sakaka, Jouf) was used to examine the films' morphology. An energy-dispersive X-ray (EDX) spectrometer (Sakaka, Jouf) connected to a thermal field emission scanning electron microscope (FESEM, Sakaka, Jouf) was used to analyze the elements with an electron gun. Electrical tests were carried out, and resistivity, hall mobility, and carrier concentration were determined at room temperature in a light magnetic field of around 0.554 T, using a hall measurement device that employed the van der Pauw method.

Ultimately, the rate at which the methylene blue (MB) aqueous solution degraded in the presence of sunlight was used to evaluate the photocatalytic activity of the structures based on the obtained thin film heterostructures. The cylindrical batch reactor used in the research was opened to the air. For the photocatalytic tests, methylene blue MB (Aldrich) was selected as the model molecule. A steady stream of water flowing through the reactor maintained the MB solution at room temperature. The starting MB concentration was 4.5 mg/L. To achieve adsorption equilibrium, the MB solution was agitated at a magnetic stirrer for one hour in the dark. Under continuous stirring, the aqueous suspension comprising MB and the photocatalysts of ZnO/α-Fe_2O_3, In_2O_3/α-Fe_2O_3, and SnO_2/α-Fe_2O_3 thin films was exposed to solar radiation. Four analytical samples were taken from the MB solution every 30 min in order to determine the impact of sunlight irradiation. Each sample's MB concentration was determined using a UV–Vis spectrophotometer.

Since the absorbance, A, and concentration, C, of MB are proportionate according to the Beer–Lambert law, the efficiency of MB degradation was determined using the following equation:

$$\text{Degradation efficiency}(\%) = \frac{A_0 - A}{A_0} \times 100 = \frac{C_0 - C}{C_0} \times 100 \quad (1)$$

where A_0, A and C_0, C represent the MB absorbance and concentration, respectively, corresponding to the start time and the variable time. Lastly, counting forming unity (CFU) was used to assess the bactericidal qualities of ZnO/α-Fe_2O_3, In_2O_3/α-Fe_2O_3, and SnO_2/α-Fe_2O_3 thin film heterostructures. The experimental strains of the Gram-negative bacteria Pseudomonas aeruginosa and Gram-positive bacteria Bacillus subtilis were employed. In Tryptic Lauria Bertani (LB) medium, the bacterial cells were cultivated at 37 °C until they reached an optical density (OD) of 0.2 at 600 nm. During the exponential growth phase, the culture was inoculated with 10^6 CFU in a fresh LB medium, and 10 μL was placed onto the $ZnO/\alpha\text{-}Fe_2O_3$, $In_2O_3/\alpha\text{-}Fe_2O_3$, and $SnO_2/\alpha\text{-}Fe_2O_3$ thin films. It was then incubated at 37 °C for 24 h in a dark environment. An identical process was applied to inert glasses devoid of thin coatings, which served as the control. Following incubation, bacterial suspensions were serially diluted and plated for CFU counting on Plate Count Agar (PCA) plates measuring 9 cm in diameter. The effect of thin films on bacterial growth was determined by looking at the decrease in CFU/mL. The residual bacterial viability was determined as follows:

$$\%\ \text{viability} = \frac{\text{CFU with thin films}}{\text{CFU of control}} \times 100 \quad (2)$$

3. Results and Discussions

3.1. Structural Properties

XRD was used to evaluate the crystal phase of the prepared $ZnO/\alpha\text{-}Fe_2O_3$, $In_2O_3/\alpha\text{-}Fe_2O_3$, and $SnO_2/\alpha\text{-}Fe_2O_3$ thin film heterostructures. Firstly, all peaks observed in the crystal structure of $ZnO/\alpha\text{-}Fe_2O_3$ (Figure 1a) were located at 2θ values of 24.3°, 31.4°, 33.4°, 34.9°, 40.6°, 44.5°, 54.2°, and 57.7°, which are related to the (012), (100), (104), (110), (113), (024), (110), and (122) planes, respectively. These peaks indicate the hexagonal phase of ZnO and the rhombohedral phase of the hematite $\alpha\text{-}Fe_2O_3$ according to the JCPDS card numbers 036-1451 and 01-1053, respectively. These findings are consistent with those of Noukelag et al., who synthesized zincite ZnO and hematite $\alpha\text{-}Fe_2O_3$ for the first time from an aqueous extract of rosemary [28].

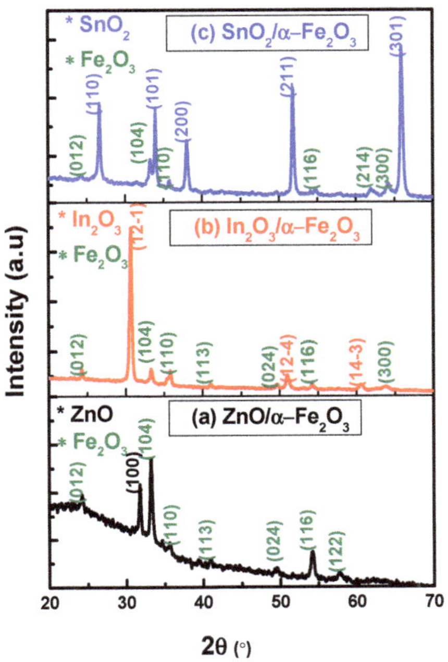

Figure 1. XRD results of (**a**) $ZnO/\alpha\text{-}Fe_2O_3$, (**b**) $In_2O_3/\alpha\text{-}Fe_2O_3$, and (**c**) $SnO_2/\alpha\text{-}Fe_2O_3$ thin film heterostructures.

The XRD of In_2O_3/α-Fe_2O_3 is shown in Figure 1b, with the primary diffraction peaks corresponding to the cubic phase of In_2O_3 (JCPDS No: 06-0416) and the rhombohedral phase of α-Fe_2O_3 (JCPDS No: 01-1053). The absence of further impurity peaks or mixed oxides suggests that the different metal oxides did not chemically interact with each other [29].

The XRD pattern of the SnO_2/α-Fe_2O_3 thin film is shown in Figure 1c. The patterns reveal many peaks at (110), (101), (200), (211), and (301), which match the lattice planes of the tetragonal structure JCPDS (no. 41–1445). In addition, weak peaks are revealed at (012), (104), (110), (116), (214), and (300) that are compatible with the JCPDS (no. 01-1053) associated with rhombohedral hematite α-Fe_2O_3 [30].

3.2. Raman Measurements

Raman spectroscopy is a dependable method for characterizing materials. It distinguishes various material phases using discrete vibrational modes. The phase type and purity of the material are confirmed using the numerous unique phonon modes revealed in the Raman spectra. Figure 2 shows the Raman spectrum analysis of the ZnO/α-Fe_2O_3, In_2O_3/α-Fe_2O_3, and SnO_2/α-Fe_2O_3 thin film heterostructures. The Eg modes of α-Fe_2O_3 are represented by the peaks at 243, 292, 404, and 604 cm^{-1}, whereas the A1g modes are associated with the peak at 497 cm^{-1}. These peaks appeared in all the different thin films. These results enhance the production of pure hematite α-Fe_2O_3 [10–12].

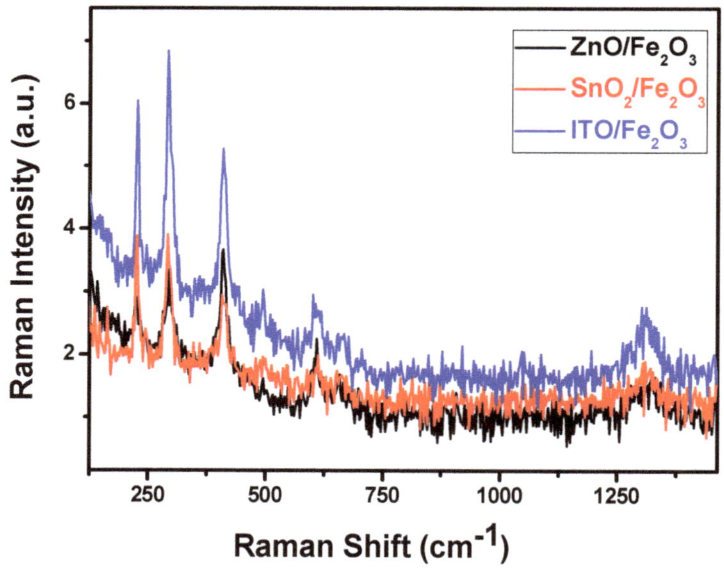

Figure 2. SEM micrographs of SnO_2/α-Fe_2O_3 thin films, In_2O_3/α-Fe_2O_3 thin films, and ZnO/α-Fe_2O_3 thin films.

3.3. Morphological Properties

SEM analysis was used to evaluate the morphology of ZnO/α-Fe_2O_3, In_2O_3/α-Fe_2O_3, and SnO_2/α-Fe_2O_3 thin film heterostructures presented in Figure 3.

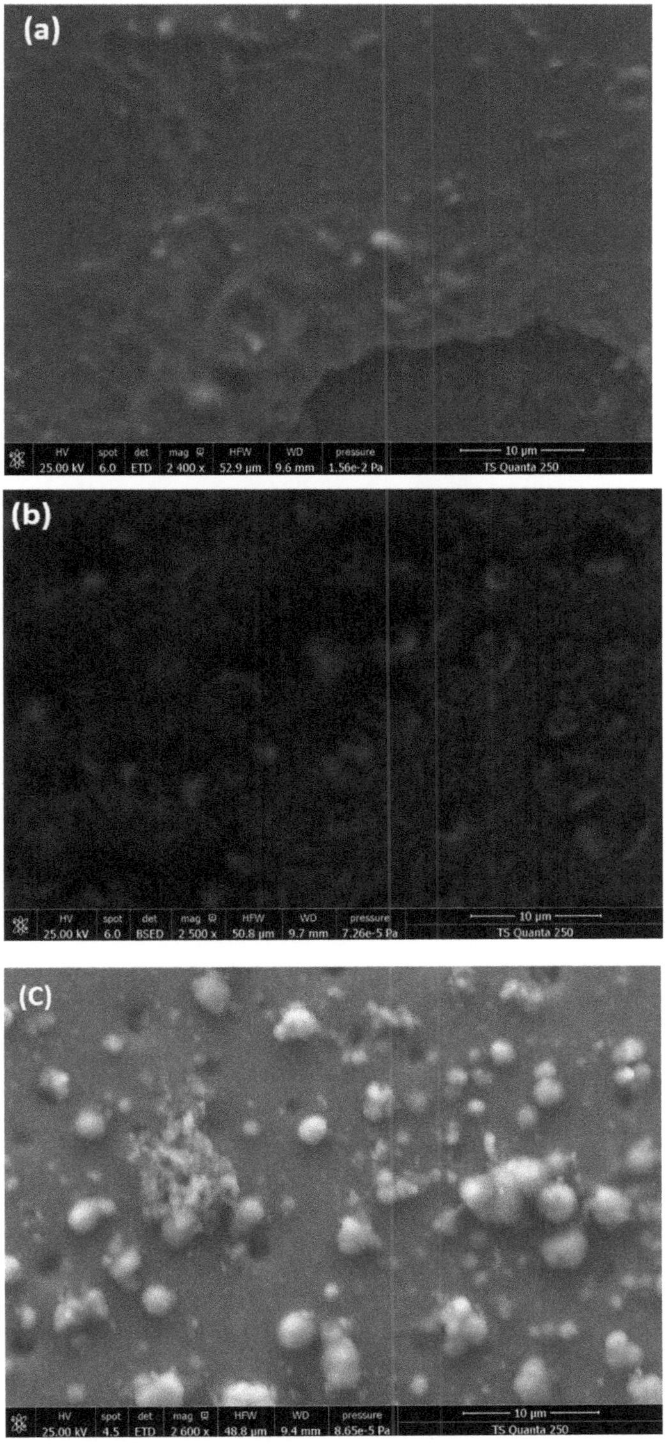

Figure 3. SEM micrographs of (**a**) $ZnO/\alpha\text{-}Fe_2O_3$ thin films, (**b**) $In_2O_3/\alpha\text{-}Fe_2O_3$ thin films, and (**c**) $SnO_2/\alpha\text{-}Fe_2O_3$ thin films.

ZnO/Fe_2O_3 has a homogeneous and smooth surface with a very small particle size. However, the thin film SEM pictures of SnO_2/Fe_2O_3 and In_2O_3/Fe_2O_3 have a completely different shape compared to ZnO/Fe_2O_3. Figure 3b,c show a highly folded, rough, uneven surface with imperfections confined between 20 and 65 nm. The elaboration process and the formation of thin layers in the heterostructure may cause these irregularities. This cracked surface improves the rapid absorption of contaminants and allows water to easily pass through the surface of our thin layers.

3.4. EDX Measurments

EDX was used to support the aforementioned findings; the atomic percentages (at.%) of Zn (from ZnO), Sn (from SnO_2), In (from In_2O_3), and Fe (from the Fe_2O_3) found on thin films are presented in Figure 4a–c, which display the corresponding mapping structure of the obtained thin films.

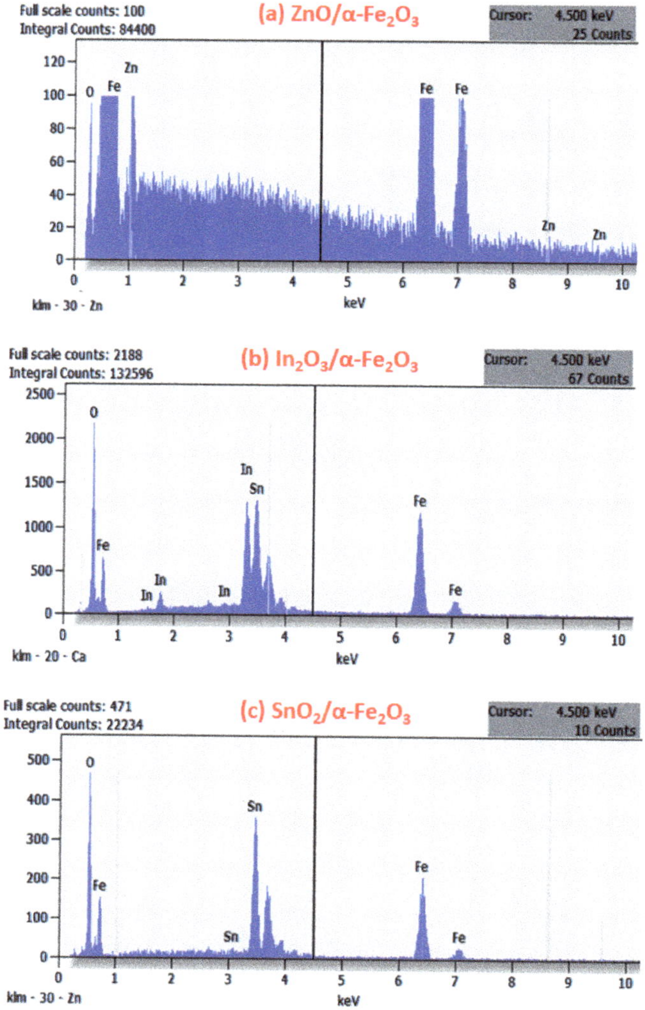

Figure 4. EDXS spectra of (**a**) ZnO/α-Fe_2O_3 thin films, (**b**) In_2O_3/α-Fe_2O_3 thin films, and (**c**) SnO_2/α-Fe_2O_3 thin films.

Only three components were found in the spectrum for these films: O, Fe, and Zn for $ZnO/\alpha\text{-}Fe_2O_3$; O, Fe, In for $In_2O_3/\alpha\text{-}Fe_2O_3$; and O, Fe, and Sn for $SnO_2/\alpha\text{-}Fe_2O_3$. No other elements were discovered within the apparatus' sensitivity range, which shows the great purity of the thin films.

On the other hand, we noted that the distribution of Fe and O signals was uniform throughout the entire structure. Seldom were the Zn, In, and Sn signals found outside of the core region; they were mostly found inside. The unique characteristics of the $ZnO/\alpha\text{-}Fe_2O_3$, $In_2O_3/\alpha\text{-}Fe_2O_3$, and $SnO_2/\alpha\text{-}Fe_2O_3$ heterostructure nanocomposites were validated by these results, which align with the findings documented in the previous literature [13,14].

3.5. Hall Effect Study

The Hall effect measurements provide important information, such as the type of material (***n***-type or ***p***-type), carrier concentration, electrical resistivity, and Hall mobility. These findings are detailed in Table 1.

Table 1. Hall effect results of $In_2O_3/\alpha\text{-}Fe_2O_3$ and $SnO_2/\alpha\text{-}Fe_2O_3$ thin films.

	Conductivity Type	Carrier Concentration (cm^{-3})	Resistivity ρ (Ω.cm)	Mobility μ ($cm^2/V.S$)
$In_2O_3/\alpha\text{-}Fe_2O_3$	n	-1.87×10^{21}	5.71×10^{-4}	5.82
$SnO_2/\alpha\text{-}Fe_2O_3$	n	-1.88×10^{20}	1.49×10^{-3}	22.20

Firstly, the results indicate that the conductivity of the $ZnO/\alpha\text{-}Fe_2O_3$, $In_2O_3/\alpha\text{-}Fe_2O_3$, and $SnO_2/\alpha\text{-}Fe_2O_3$ thin films was ***n***-type in nature and all the information shows notable changes between the obtained thin film heterostructures. The increase in the carrier concentration and the mobility of $SnO_2/\alpha\text{-}Fe_2O_3$ can be attributed to the increase in the number of the grain boundary on the surface of layer and the corresponding increase in the dislocation density across the boundaries.

On the other hand, the heterostructure of the $ZnO/\alpha\text{-}Fe_2O_3$ thin film has a high resistivity. Consequently, the results showed no changes.

3.6. Photocatalytic Test

Adsorption is a crucial step in every catalytic reaction process, regardless of the type of photodegradation employed. More effective adsorption enables improved pollutant–catalyst interactions. In this work, MB was adsorbed on the synthesized thin film heterostructures at room temperature, as plotted in Figure 5.

During photoelectrochemical water-splitting, the observed photocurrent is adversely affected by the consumption of photoexcited electrons. Under sun irradiation, $ZnO/\alpha\text{-}Fe_2O_3$, $In_2O_3/\alpha\text{-}Fe_2O_3$, and $SnO_2/\alpha\text{-}Fe_2O_3$ are excited. The photogenerated electrons quickly move from ZnO, In_2O_3, and SnO_2 CB to $\alpha\text{-}Fe_2O_3$'s VB, allowing for holes to collect in ZnO, In_2O_3, and SnO_2 VB, and for electrons to gather in high-potential $\alpha\text{-}Fe_2O_3$ CB, where they can reduce H^+ in H_2.

Morphology becomes important at this point. The photocatalyst's precise engineering and design may be able to solve the issues with charge carrier transit and recombination. Indeed, there are numerous indications of the existence of a relationship between reduced-dimension structures and enhanced photocurrent density, which is linked to the occurrence of defects in nanostructures. Under visible light irradiation, we discovered that the concentration of MB decreases and the peak almost vanishes after six hours, but only for $SnO_2/\alpha\text{-}Fe_2O_3$. This good response is due to the presence of a cracked surface, which increases the rapid absorption of contaminants and allows water to easily pass through the surface of our thin layers. Because of its improved nanoarchitecture, which facilitates dye molecule diffusion and oxygen species transport during the photoactivity process, the porous sample exhibits a higher photocatalytic performance than the pure one. Electron–hole pairs (e^-, h^+) are produced when photon energy (hv) is absorbed by the thin film heterojunctions during radiation.

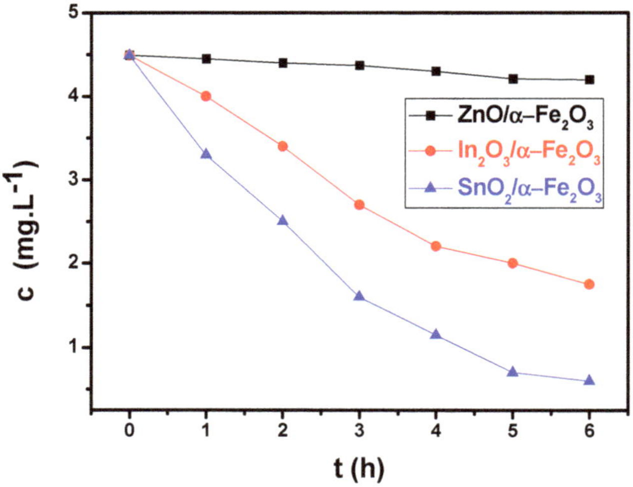

Figure 5. Photocatalysis of methylene blue: normalized concentration of MB as a function of time for $ZnO/\alpha\text{-}Fe_2O_3$, $In_2O_3/\alpha\text{-}Fe_2O_3$, and $SnO_2/\alpha\text{-}Fe_2O_3$ thin films under solar radiation.

After photogeneration occurs close to the surface, the electrons combine with oxygen to make superoxide ($^{\cdot}O_2^-$) (Equation (3)), which then reacts with water to form hydroxyl radicals ($^{\cdot}OH$). In parallel with this, the water and photogenerated holes react to produce $^{\cdot}OH$ radicles.

Intermediate reactions also result in the formation of $^{\cdot}OH$ radicals and the powerful oxidants finally convert MB into CO_2 and H_2O.

$$O_2 + e^- \rightarrow {}^{\cdot}O_2^- \quad (3)$$

$${}^{\cdot}O_2^- + H_2O \rightarrow OH^- + {}^{\cdot}HO_2 \quad (4)$$

$$H_2O + h^+ \rightarrow {}^{\cdot}OH + H^+ \quad (5)$$

$$h^+ + OH^- \rightarrow {}^{\cdot}OH \quad (6)$$

$${}^{\cdot}HO_2 + H_2O \rightarrow H_2O_2 + {}^{\cdot}OH \quad (7)$$

$$H_2O_2 + e^- \rightarrow {}^{\cdot}OH + OH^- \quad (8)$$

$$MB + ({}^{\cdot}OH + {}^{\cdot}O_2^-) \rightarrow CO_2 + H_2O \quad (9)$$

3.7. Antibacterial Activity

The antibacterial effectiveness against Gram-negative Pseudomonas aeruginosa (***Pa***) and Gram-positive Bacillus subtilis (***Bs***) bacteria was also assessed using $ZnO/\alpha\text{-}Fe_2O_3$, $In_2O_3/\alpha\text{-}Fe_2O_3$, and $SnO_2/\alpha\text{-}Fe_2O_3$ thin films. The $In_2O_3/\alpha\text{-}Fe_2O_3$ thin film demonstrated a residual viability equal to 0.5% and 0.187% of bacterial cells against of ***Pa*** and of ***Bs***, respectively. In addition, the $SnO_2/\alpha\text{-}Fe_2O_3$ showed a viability of 0.65% and 0.62%. However, the results of the bacteria test in the presence of $ZnO/\alpha\text{-}Fe_2O_3$ revealed higher values, as illustrated in Table 2. The generation of reactive oxygen species (ROS) and the electrostatic interaction between $In_2O_3/\alpha\text{-}Fe_2O_3$ and the absorbed OH- and H_2O may be the cause of the antibacterial action. One possible explanation for the increased antibacterial properties of $In_2O_3/\alpha\text{-}Fe_2O_3$ thin films could be their greater root mean square roughness (Rms) in relation to that of the other thin films.

Table 2. Antibacterial activity of $In_2O_3/\alpha\text{-}Fe_2O_3$, $SnO_2/\alpha\text{-}Fe_2O_3$, and $ZnO/\alpha\text{-}Fe_2O_3$ thin films against Pseudomonas aeruginosa and Bacillus subtilis bacteria.

		Pseudomonas Aeruginosa		**Bacillus Subtilis**	
			% Viability		**% Viability**
Bacteria without sample		2×10^8	--	4×10^7	--
Bacteria in the presence of	$In_2O_3/\alpha\text{-}Fe_2O_3$	10^6	0.5	7.5×10^4	0.187
	$SnO_2/\alpha\text{-}Fe_2O_3$	2×10^6	0.65	2.5×10^5	0.625
	$ZnO/\alpha\text{-}Fe_2O_3$	1.3×10^6	1	4×10^6	10

The combined benefits of the photocatalytic activity and the antibacterial impact are crucial for a variety of applications (medical, food storage, etc.). To the best of our knowledge, thin films have never been used to deteriorate ***Pa*** and ***Bs***.

4. Conclusions

This study presents the preparation of heterostructure $ZnO/\alpha\text{-}Fe_2O_3$, $In_2O_3/\alpha\text{-}Fe_2O_3$, and $SnO_2/\alpha\text{-}Fe_2O_3$ thin films using the spray pyrolysis method. The XRD and Raman spectroscopy confirm the presence of a rhombohedral structure of $\alpha\text{-}Fe_2O_3$ and hexagonal phase of ZnO, a cubic phase of In_2O_3, and a tetragonal structure of SnO_2. The surface morphology of the obtained thin films was analyzed via scanning electron microscopy (SEM). The results show a homogeneous and smooth surface with a very smaller particle size for ZnO/Fe_2O_3. Contrarily, the thin film SEM pictures of SnO_2/Fe_2O_3 and In_2O_3/Fe_2O_3 show a highly folded, rough, uneven surface with imperfections at 20 and 65 nm. The Hall effect indicates an increase in the carrier concentration and the mobility of $SnO_2/\alpha\text{-}Fe_2O_3$, which can be attributed to the increase in the number of grain boundaries on the surface of the layer and corresponding increase in the dislocation density across the boundaries. When the photocatalytic activity of the sprayed thin films was evaluated, it was found that the paired $In_2O_3/\alpha\text{-}Fe_2O_3$ photoatalysts had an impact on the photocatalytic activity and degradation mechanism of the MB dye. This fractured surface enhances the speed at which impurities are absorbed and facilitates water's easy penetration into the surface of the thin layers. Additionally, when a thin coating of $In_2O_3/\alpha\text{-}Fe_2O_3$ was used, only 5% of Gram-negative Pseudomonas aeruginosa (***Pa***) and 0.187% of Gram-positive Bacillus subtilis (***Bs***) bacterial cells were still viable. As this heterostructure can be produced using an easy spray pyrolysis approach, these results are very interesting. This study offers a path for additional research in the healthcare industry. The In_2O_3/α-Fe_2O_3 and SnO_2/α-Fe_2O_3 thin films yielded the most effective combined efficiency.

Author Contributions: This paper was edited equally by A.A. and A.M. All authors have read and agreed to the published version of the manuscript.

Funding: This work was funded by the Deanship of Graduate Studies and Scientific Research at Jouf University under grant No (DGSSR-2023-02-02337).

Data Availability Statement: We declare that all data are available in the manuscript file.

Acknowledgments: We declare that this manuscript is original and has not been published elsewhere, nor is it currently under consideration for publication elsewhere.

Conflicts of Interest: The authors declare no conflict of interest.

References

1. Rzaij, J.M. A novel room-temperature nitrogen dioxide gas sensor based on silver-doped cerium oxide thin film. *Sens. Actuatots A Phys.* **2023**, *363*, 114748. [CrossRef]
2. Rathod, A.P.S.; Mishra, P.K.; Mishra, A. Fast-response/recovery In_2O_3 thin-film transistor-type NO_2 gas sensor with floating-gate at low temperature. *Sens. Actuators B Chem.* **2023**, *394*, 134477.
3. Zhao, T.; Ye, Y.; Guo, K.; Cui, R.; Zhang, M.; Wang, X.; Zhang, B.; Zhang, J.; Deng, C. High energy storage properties of calcium-doped barium titanate thin films with high breakdown field strength. *J. Alloys Compd.* **2024**, *970*, 172487. [CrossRef]
4. Song, H.; Son, J.Y. Energy storage and mutiferroic properties of La-doped epitaxial $BiFeO_3$ thin films according to La doping concentration. *J. Energy Storage* **2023**, *68*, 107729. [CrossRef]
5. Wang, X.; Chen, J.; Hu, S.; Yu, K.; Yang, F.; Shi, Y.; Li, J.; Hou, M.; Liu, A.; Zheng, M.; et al. Enhanced energy storage properties in $PbZrO_3$ thin films via the incorporation of NiO. *Curr. Appl. Phys.* **2023**, *52*, 24. [CrossRef]
6. Cantos, A.; Gundogan, S.H.; Turkoglu, F.; Koseoglu, H.; Aygun, G.; Ozyuzer, L. Photovoltaic performance of magnetron sputtered antimony selenide thin film solar cells buffered by cadmium sulfide and cadmium sulfide/zinc sulfide. *Thin Solid Films* **2023**, *784*, 140070. [CrossRef]
7. Franco, M.A.M.; Castaneda, C.A.R.; Romero, P.M.M.; Barragan, J.J.P.; Quintero, O.A.J.; Hu, H. A direct correlation between structural and morphological defects of TiO_2 thin films on FTO substrates and photovoltaic performance of planar perovskite solar cells. *Mater. Sci. Semicond. Process.* **2023**, *161*, 107452. [CrossRef]
8. Akcay, N.; Gremenok, V.F.; Ozen, Y.; Buskis, K.P.; Zaretskaya, E.P.; Ozcelik, S. Investigation on photovoltaic performance of Cu_2SnS_3 thin films solar cells fabricated by PF-sputtered In_2S_3 buffer layer. *J. Alloys Compd.* **2023**, *949*, 169874. [CrossRef]
9. Yang, F.; Wang, P.; Hao, J.; Qu, J.; Cai, Y.; Yang, X.; Li, C.M.; Hu, J. Ultrasound-assisted piezoelectric photocatalysis: An effective strategy for enhancing hydrogen evolution from water splitting. *Nano Energy* **2023**, *118*, 108993. [CrossRef]
10. Sarma, M.; Jaiswal, M.K.; Podder, S.; Bora, J.; Karmakar, S.; Choudhury, B.; Pal, A.R. A study on the applicability of thin film over powder for visible light photocatalysis. *Phys. B Condens. Matter* **2023**, *670*, 415354. [CrossRef]
11. Li, A.; Zhang, P.; Kan, E.; Gong, J. Wettability adjustment to enhance mass transfer for heterogeneous electrocatalysis and photocatlysis. *eScience* **2023**, *4*, 100157. [CrossRef]
12. Muffer, H.J.; Fischer, C.H.; Diesner, K.; Steiner, M.C.L. ILGAR- A novel film technology for sulfides. *Sol. Energy Mater. Sol. Cells* **2001**, *67*, 121–127. [CrossRef]
13. Sati, D.C.; Dahshan, A.; Hegazy, H.H.; Aly, K.A.; Sharma, P. Optical and optoelectronic properties of (Ge2S8)100-xTex thin films for IR optical device fabrication. *Surf. Interfaces* **2023**, *39*, 102995. [CrossRef]
14. Timoumi, A.; Alameer, O.; Alamri, S. Intensive study of coating multilayer TiO_2 nanoparticles thin films used for optoelectronics devices. *Results Mater.* **2023**, *18*, 100390. [CrossRef]
15. Ahmed, M.; Bakry, A.; Shaaban, E.R. Optical characteristics with high accuracy of diluted Cr doped In2O3 thin films using spectroscopic ellipsometry for optoelectronic devices. *Opt. Mater.* **2022**, *133*, 113039. [CrossRef]
16. Prakash, P.; Janarthanan, B. Enhancement of sensitization and electron transfer by kumkum dye in dye in dye sensitized solar cell applications. *Optik* **2023**, *287*, 171093. [CrossRef]
17. Huang, Y.-T.; Ono, T. Magnetostriction of electroplated TbFeCo thin films. *J. Magn. Magn. Mater.* **2023**, *577*, 170799. [CrossRef]
18. Semnani, S.Z.M.; Youselpour, M.; Zareidoost, A. Enhancing the biocompatibility of ZrO_2 thin film on Zr-2.5Nb alloy by anodizing treatment using an electrolyte containing biofunctional groups. *Thin Solid Films* **2022**, *753*, 139279. [CrossRef]

9. Nishii, H.; Iida, S.; Yamasaki, A.; Ikenoue, T.; Miyake, M.; Doi, T.; Hirato, T. Fabrication of epitaxial V_2O_3 thin films on Al_2O_3 substrates via mist chemical vapor deposition. *J. Cryst. Growth* **2024**, *626*, 127484. [CrossRef]
0. Lin, B.; Cai, Y.; Wang, Y.; Zou, Y.; Gao, C.; Liu, Y.; Liu, W.; Guo, S.; Sun, C. Effects of growth temperature and reactor pressure on AlN thin film grown by metal-organic chemical vapor deposition. *Thin Solid Films* **2023**, *783*, 140037. [CrossRef]
1. Khan, R.; Ali-Löytty, H.; Tukiainen, A.; Tkachenko, N.V. Comparison of the heat-treatment effect on carrier dynamics in TiO_2 thin films deposited by different methods. *Phys. Chem. Chem. Phys.* **2021**, *23*, 17672. [CrossRef] [PubMed]
2. Saari, J.; Ali-Loytty, H.; Kauppinen, M.M.; Hannula, M.; Khan, R.; Lahtonen, K.; Palmolahti, L.; Tukiainen, A.; Gronbeck, H.; Tkachenko, N.V.; et al. Tunable Ti^{3+}-Mediated Charge Carrier Dynamics of Atomic Layer Deposition-Grown Amorphous TiO_2. *J. Phys. Chem. C* **2022**, *126*, 4542–4554. [CrossRef]
3. Zhang, W.; Zhao, X.; Sun, Q.; Gao, W.; Yang, X.; Wang, X. Optical, electrical and thermal stability properties of Al and F co-doped ZnO thin films prepared by sol-gel spin-coating. *Thin Solid Films* **2023**, *776*, 139889. [CrossRef]
4. Alofi, S.; O'Rourke, C.; Mills, A. Photocatalytic destruction of stearic acid by TiO2 films: Evidence of highly efficient transport of photogenerated electrons and holes. *J. Photochem. Photobiol. A Chem.* **2023**, *435*, 114273. [CrossRef]
5. Singh, J.; Rathi, A.; Rawat, M.; Kumar, V.; Kim, K.H. The effect of manganese doping on structural, optical, and photocatalytic activity of zinc oxide nanoparticles. *Compos. Parts B Eng.* **2019**, *166*, 361–370. [CrossRef]
6. Rana, S.; Kumar, A.; Dhiman, P.; Sharma, G.; Amirian, J.; Stadler, F.J. Progress in graphdiyne and phosphorene based composites and heterostructures as new age materials for photocatalytic hydrogen evolution. *Fuel* **2024**, *356*, 129630. [CrossRef]
27. Mhamdi, A.; Alkhalifah, M.S.; Rajeh, S.; Labidi, A.; Amlouk, M.; Belgacem, S. Electrical and gas sensing investigations on the sprayed ZnO:Cu thin films. *Phys. B* **2017**, *521*, 178. [CrossRef]
28. Noukelag, S.K.; Gummings, F.; Arendse, C.J.; Maaza, M. Physical and magnetic properties of biosynthesized ZnO/Fe_2O_3, ZnO/$ZnFe_2O_4$ and $ZnFe_2O_4$ nanoparticles. *Results Surf. Interfaces* **2023**, *10*, 100092. [CrossRef]
29. Ivanovskaya, M.; Kotsikau, D.; Faglia, G.; Nelli, P. Influence of chemical composition factors of Fe_2O_3/In_2O_3 sensors on their selectivity and sensitivity to ethanol. *Sens. Actuators B* **2003**, *96*, 498–503. [CrossRef]
30. Yuan, T.; Jiang, Y.; Li, Y.; Zhang, D.; Yan, M. Enhanced lithium storage performance in three-dimensional porous SnO_2-Fe_2O_3 composite anode films. *Electrochimica Acta* **2014**, *136*, 27–32. [CrossRef]

surfaces

MDPI

Article

Highly Efficient Cobalt Sulfide Heterostructures Fabricated on Nickel Foam Electrodes for Oxygen Evolution Reaction in Alkaline Water Electrolysis Cells

Ioannis Poimenidis [1,*], Nikandra Papakosta [2,3], Panagiotis A. Loukakos [2], George E. Marnellos [4,5] and Michalis Konsolakis [1,*]

1 Lab of Matter Structure and Laser Physics, School of Production Engineering & Management, Technical University of Crete, 73100 Chania, Crete, Greece
2 Foundation for Research and Technology–Hellas, Institute of Electronic Structure and Laser, 70013 Heraklion, Greece; papakosta@iesl.forth.gr (N.P.); loukakos@iesl.forth.gr (P.A.L.)
3 Department of Materials Science and Technology, University of Crete, Vassilika Vouton, 70013 Heraklion, Greece
4 Department of Chemical Engineering, Aristotle University of Thessaloniki, 54124 Thessaloniki, Greece; gmarnellos@cheng.auth.gr or
5 Chemical Process and Energy Resources Institute, Centre for Research & Technology Hellas, 57001 Thessaloniki, Greece
* Correspondence: ipoimenidis@tuc.gr (I.P.); mkonsolakis@tuc.gr (M.K.)

Abstract: Non-noble metal electrocatalysts for the oxygen evolution reaction (OER) have recently gained particular attention. In the present work, a facile one-step electrodeposition method is applied in situ to synthesize cobalt sulfide nanostructures on nickel foam (NF) electrodes. For the first time, a systematic study is carried out on the impact of the Co/S molar ratio on the structural, morphological, and electrochemical characteristics of Ni-based OER electrodes by employing $Co(NO_3)_2{\cdot}6\ H_2O$ and CH_4N_2S as Co and S precursors, respectively. The optimum performance was obtained for an equimolar Co:S ratio (1:1), whereas sulfur-rich or Co-rich electrodes resulted in an inferior behavior. In particular, the Co_xS_y@NF electrode with Co/S (1:1) exhibited the lowest overpotential value at 10 mA cm^{-2} (0.28 V) and a Tafel slope of 95 mV dec^{-1}, offering, in addition, a high double-layer capacitance (C_{DL}) of 10.7 mF cm^{-2}. Electrochemical impedance spectroscopy (EIS) measurements confirmed the crucial effect of the Co/S ratio on the charge-transfer reaction rate, which is maximized for a Co:S molar ratio of 1:1. Moreover, field emission scanning electron microscopy (FE-SEM), X-ray diffraction (XRD) and X-ray fluorescence (XRF) were conducted to gain insights into the impact of the Co/S ratio on the structural and morphological characteristics of the electrodes. Notably, the Co_xS_y@NF electrocatalyst with an equimolar Co:S ratio presented a 3D flower-like nanosheet morphology, offering an increased electrochemically active surface area (ESCA) and improved OER kinetics.

Keywords: alkaline electrolysis; oxygen evolution reaction; electrodeposition on Ni foam; cobalt–sulfur molar ratio; thiourea

1. Introduction

The fossil fuel crisis and global climate change have led researchers to pay increased attention to alternative energy sources. Hydrogen has great potential to be employed as an energy carrier for the forthcoming energy transition due to its high gravimetric energy density and zero carbon content [1–4]. In this regard, the electrochemical splitting of water using energy derived from intermittent renewable energy sources (green Hydrogen), such as photocatalysis, solar thermochemical, photovoltaic electrolysis, and the supercritical water gasification of biomass, has recently gained particular importance [5].

The maturity of the electrochemical water splitting process will highly contribute to the need for green hydrogen production in the years to come, which, according to the International Energy Agency analysis, will be ~150 Mt and ~435 Mt of low-carbon hydrogen production in 2030 and 2045, respectively [5].

However, it is well known that the efficiency of water electrolysis is associated with the overpotentials of the hydrogen evolution reaction (HER) and oxygen evolution reaction (OER) [6,7]. The high overpotential of the OER ($4OH^- \rightarrow 2H_2O + 4e^- + O_2$ in alkaline medium), due to the demanding transfer of four electrons in the anodic charge transfer reaction, represents the main bottleneck for the limited efficiencies [8–11]. The state-of-the-art electrodes used in the OER mainly comprise Ir and Ru metals; however, their scarcity and high cost notably hinder their widespread application as electrocatalysts [12–14].

Significant research has recently been devoted to noble-metal-free and earth-abundant transition metal electrocatalysts [15]. Various materials, such as borides, chalcogenides, etc., have been explored for the OER. However, most of these candidates suffer from a small active electrochemical area and low electrical conductivity, negatively affecting OER kinetics [15]. Recently, the in situ growth of nanostructured catalytic materials on conductive substrates, such as nickel foam (NF), has proven an efficient approach to overcoming these obstacles [15,16].

In recent years, transition-metal-based electrocatalysts have been developed for water splitting via various techniques, such as aerosol spray [17], photochemical metal–organic deposition [18], hydrothermal processes [19–22], pulsed laser deposition [23], etc. Among them, transition metal sulfides (TMSs) have been considered excellent alternatives for the OER due to their low cost, adjustable electronic properties, and adequate conductivity, which make them suitable alternatives to the Ru and IrO_2 benchmark electrodes [24–26].

Although various studies have been conducted using TMSs, especially with Co_xS_y composites, there is no systematic study on the impact of the sulfur/cobalt molar ratio on the electrochemical performance of electrocatalysts. Nonetheless, the type and concentration of the precursor compounds employed during fabrication are expected to affect the structural, morphological, and electrochemical properties of the electrocatalysts. Furthermore, in most studies, the fabricated electrocatalysts are in the form of powders, rendering the use of binders for the electrocatalyst's anchor to the substate unavoidable, thus increasing the contact resistance [16,27]. In this regard, developing a facile and binder-free preparation protocol for highly efficient OER electrodes is paramount.

Motivated by the above challenges, the present work aims to investigate the effect of the sulfur/cobalt ratio on the OER. Although several studies have been devoted to cobalt sulfide electrocatalysts, there is no systematic study on the impact of sulfur/cobalt on their physicochemical properties and, in turn, on their OER performance. In this regard, a facile one-step electrodeposition method was used to fabricate Co_xS_y (x:y molar ratio) binary electrodes on a nickel foam substrate (NF), employing thiourea (CH_4N_2S) and cobalt nitrate hexahydrate ($Co(NO_3)_2{\cdot}6\ H_2O$) as precursors. Various Co:S molar ratios were used during the electrodeposition process, and it was clearly disclosed that the Co:S ratio significantly affects the morphological and structural properties of the electrode and its electrochemical performance. Remarkably, the Co_xS_y@NF electrode with an equimolar Co:S ratio (1:1) presented the lowest overpotential values at 10 mA cm^{-2} (0.28 V), the lowest Tafel slope (95 mV dec^{-1}), and the highest C_{DL} (10.7 mF cm^{-2}) and ECSA (537 cm^2) values. On the other hand, sulfur-rich or Co-rich electrodes resulted in an inferior behavior, demonstrating the key effect of sulfur content on transition metal sulfide electrocatalysts.

2. Materials and Methods

2.1. Materials and Reagents

The chemical reagents in this work were used as received. Thiourea (98% Penta Chemical, Czech Republic-Prague), KOH (technical grade, Sigma-Aldrich, USA-VT-Burlington), ethanol (99.8%, ACROS Organics, Belgium-Antwerp), HCl (98% Sigma Aldrich), $Co(NO_3)_2{\cdot}6$

H_2O (Sigma-Aldrich) and nickel foam (99.8%, Beike advanced materials Store, China-Taizhou) were used as the electrodeposition substrate.

2.2. Electrodeposition on NF Substrate

The electrodeposition process was carried out by using an electrochemical station (Princeton Applied Research, USA-TN-Oak Ridge, VersaSTAT 4) equipped with a standard three-electrode electrochemical cell (Palmsens, Netherlands-Utrecht). The reference electrode was an Ag/AgCl electrode (3.5M KCl) (Palmsens), the counter electrode was a Pt wire (99% Goodfellow, USA-UT-Lindon), and the working electrode was an NF substrate.

Prior to the electrodeposition, the NF substrates were ultrasonicated in 3 M HCl for 10 min to activate the substrate and remove any organic impurities. Then, it was rinsed with deionized water, ultrasonicated again for 10 min in pure ethanol, and finally rinsed with deionized water.

The electrodeposition for all the samples lasted 10 min at 298 K, and the applied potential was −0.18 V vs. RHE. For the uniform deposition of the nanomaterials, a magnetic stirrer was used. The molar ratio of the $Co(NO_3)_2 \cdot 6\,H_2O$ and thiourea (CH_4N_2S) precursors during the synthesis procedure was varied in a wide range (i.e., 0:1, 1:0, 1:2, 1:1, 2:1, 4:1) to obtain both sulfur-rich and Co-rich electrodes. The as-prepared electrodes were defined as Co:S (x:y), where x:y represents the Co:S molar ratio of the CoS_x@NF composites, whereas the bare nickel foam substrate was designated as NF. Pure cobalt (Co:S (1:0)) and sulfur (Co:S (0:1)) electrodes on the NF substrate (1 × 1 cm) were also fabricated for comparison purposes. After electrodeposition, a calcination procedure was applied (at 250 °C in an air atmosphere) in order to increase the material's crystallinity and remove any surface impurities.

2.3. Structural and Morphological Characterization

The morphological characterization of the fabricated electrodes was carried out with field emission scanning electron microscopy (FE-SEM, JSM7000F, JEOL) at various magnification scales. Also, XRD (BEDE D1 with CuKa radiation) and XRF (Amptek X-123) analyses were conducted to gain insights into the structural features of the Co_xS_y@NF electrodes.

2.4. Electrochemical Evaluation of the Fabricated Electrodes

Electrochemical tests were undertaken for each developed electrode to assess its oxygen evolution reaction kinetics and electrochemical performance under water–alkaline electrolysis conditions. Specifically, cyclic voltammetry (CV), linear sweep voltammetry (LSV), and electrochemical impedance spectroscopy (EIS) measurements were carried out, and the values of the Tafel slope, double layer capacitance (C_{DL}), and ECSA were calculated. The above studies were accomplished using the VersaSTAT 4 electrochemical workstation, which was equipped with a Pt (99%) counter electrode and an Ag/AgCl (3.5 M KCl) reference electrode, and each electrodeposited-fabricated NF electrode acted as a working electrode. The alkaline electrolyte used in all tests was 1 M KOH. Before each measurement, N_2 gas was purged into the electrochemical cell to remove any amount of dissolved oxygen gas. All the experiments were repeated at least three times to confirm the obtained results. Moreover, the obtained potential values were converted to a reversible hydrogen electrode (RHE), according to the following equation:

$$E_{RHE} = E_{Ag/AgCl} + 0.059 \cdot pH + E_{oAg/AgCl} \quad (1)$$

where $E_{Ag/AgCl}$ is the potential measured with the Ag/AgCl reference electrode.

3. Results and Discussion

3.1. Electrodeposition

The electrochemical reaction rate and diffusion process are the main factors that determine the overall electrodeposition process. Mass transport limitations control the overall process when the charge transfer reaction is fast. On the other hand, if the diffusion rate is faster than the charge transfer reaction rate, the electrochemical reaction controls the process [28].

During the electrodeposition of the precursors used in this work, the following reactions are considered [29–31]:

$$2H_2O + 2e^- \rightarrow 2OH^- + H_2 \quad (2)$$

$$2Co^{2+} + 2e^- \rightarrow Co(OH)_{2(ads)} \quad (3)$$

$$Co(OH)_{2(ads)} + 2e^- \rightarrow Co + 2OH^- \quad (4)$$

$$CH_4N_2S + 2OH^- \rightarrow S^{2-} + CH_4N_2O + H_2O \quad (5)$$

$$Co^{2+} + S^{2-} \rightarrow CoS_{x\,(ads)} \quad (6)$$

Moreover, recent studies have introduced a new synthetic route for metal sulfides via the formation of an $(NH_2)_2CS_s$-M^{2+}-OH^- complex, which decomposes to the metal sulfide according to the following process [30,32]:

$$M^{2+} + CS(NH_2)_2 + 2OH^- \rightleftarrows M(OH)_2CS(NH_2)_2 \rightarrow MS + H_2NCN + 2H_2O \quad (7)$$

3.2. Structural and Morphological Evaluation

Figure 1 presents the XRD patterns of the NF, Co:S (1:0), Co:S (0:1), Co:S (1:1), Co:S (2:1), Co:S (1:2), Co:S (4:1) electrodes. The XRD analysis revealed three major peaks for the bare NF electrode at the 2θ angles of 45.2°, 52.5°, and 77°, which correspond to the (111), (200), and (220) planes of crystalline Ni, respectively. On the Co-free electrode, i.e., the Co:S (0:1) sample, there was a shift in the major peaks to 44.7°, 52.0°, and 76.5°, which are attributed to the (102), (110), and (202) crystalline planes, respectively, of the NiS phase (JCPDS file No. 02-1280) [33,34]. For the Co:S (x:y) electrodes, containing both Co and S, three major features were identified: at 44.8°, attributed to the (400) plane of the face-centered cubic Co_3O_4 structure (JCPDS No. 73-1701); 52.10°, ascribed to the (440) plane of the Co_9S_8 (PDF No.01-086-2273); and at 76.5°, attributed to the (202) crystalline plane of NiS [34–36]. No metallic phase of Ni was detected due to the formation of NiS or due to the low content (undetectable) of Ni. The XRF spectra and EDS analysis (Figures S1 and S2) also revealed the presence of sulfur, nickel, and cobalt on the as-prepared electrodes, further verifying the successful formation of Co_xS_y heterostructures according to the procedure described below (Section 2.2). These heterostructures could be responsible for the enhanced electrical conductivity and supercapacitor performance, as further discussed below on the basis of the electrochemical impedance spectroscopy (EIS) results and in agreement with relevant studies [27]. Also, it is worth noticing that S-treatment (Co:S (0:1) did not result in any structural deformation of the NF substrate, since both samples exhibited similar lattice parameters (Table S1). However, all Co-containing samples obeyed a much higher lattice parameter, indicating the formation of new Co_xS_y phases and strain in the crystal lattice.

The mean particle size for the as-prepared electrodes was calculated using the Scherrer equation, and the obtained values are summarized in Table 1. The smallest particle size of 27.5 nm was obtained for the Co:S (1:1) electrode. The rest of the fabricated electrodes presented higher crystallite sizes, varying between 28 and 35 nm. Interestingly, the optimum electrochemical behavior (see below) was obtained for the equimolar Co:S (1:1) electrode, probably implying a structure–performance relationship; a smaller particle size could result in an extended active electrochemical zone, facilitating the oxygen evolution reaction, in accordance with relevant studies [37]. This is further discussed below, based on electrochemical and impedance spectroscopy studies.

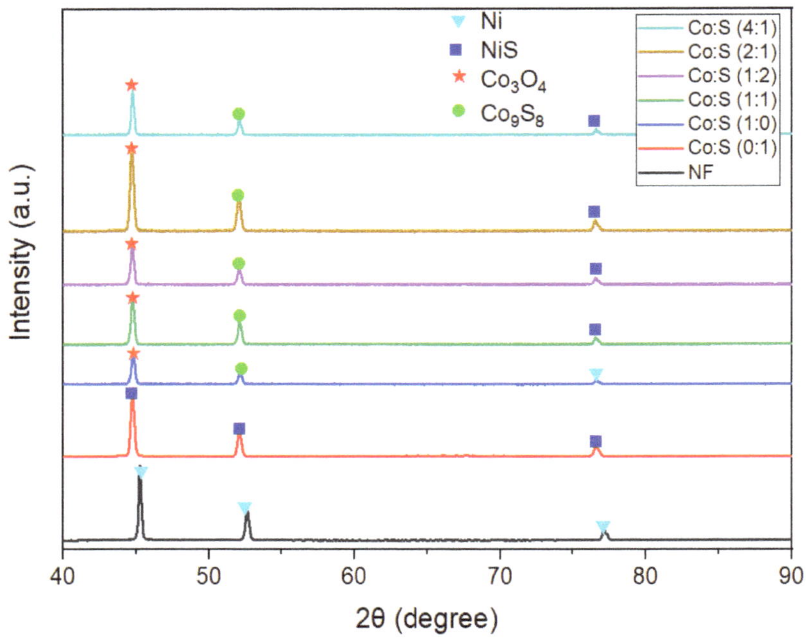

Figure 1. XRD patterns of the bare NF and Co_xS_y@NF electrodes with different Co:S molar ratios.

Table 1. Mean particle size for the NF and Co_xS_y@ NF electrodes with different Co:S molar ratios.

Electrode	Mean Particle Size (nm)
NF	35.3
Co:S (1:0)	28.1
Co:S (0:1)	31.5
Co:S (1:2)	30.1
Co:S (2:1)	29.2
Co:S (4:1)	32.2
Co:S (1:1)	27.5

In Figure 2, the FE-SEM morphological analysis of the Co_xS_y@NF electrodes is presented. In the S-only electrode (Co:S (0:1)), no obvious modifications compared to the bare NF substrate were obtained (Figure 2). In the Co-rich electrodes (Co:S (2:1) and Co:S (4:1) samples), nanosheets of irregular morphology with dense agglomerations were observed (Figure 2a,b). Similarly, where the sulfur was in excess (Figure 2c), bulky structures were obtained without an apparent formation of nanosheets (Figure 2c). Efficient electro-catalysis requires an easy flow of the involved neutral and charged species toward and from the catalytic sites; thus, the above morphology, with the presence of agglomerates, corroborates well with the inferior performance of the Co-rich and S-rich electrodes for the OER (see below) [15].

Interestingly, in the case of Co_xS_y@NF electrodes with an equimolar Co:S ratio (Figure 2d–f), the formation of well-tuned 3D flower-like nanosheets was observed. This distinct morphology is most evident in higher magnification images (Figure 2e,f), where the formation of flower-like nanosheets with a vertical orientation and a thickness of ca. 7.0 nm can be observed. These findings imply the crucial effect of the Co:S ratio on the structural and morphological characteristics of the as-prepared Co_xS_y@NF electrodes, which are expected to affect the overall electrochemical performance. In other words, only the equimolar Co:S ratio results in a distinct nano-architecture of 3D flower-like nanosheets without creating agglomerates, which can notably enhance the ESCA; the as-formed in-plain pores may bring about additional active sites, allowing a better electrolyte flow and faster OER kinetics [37].

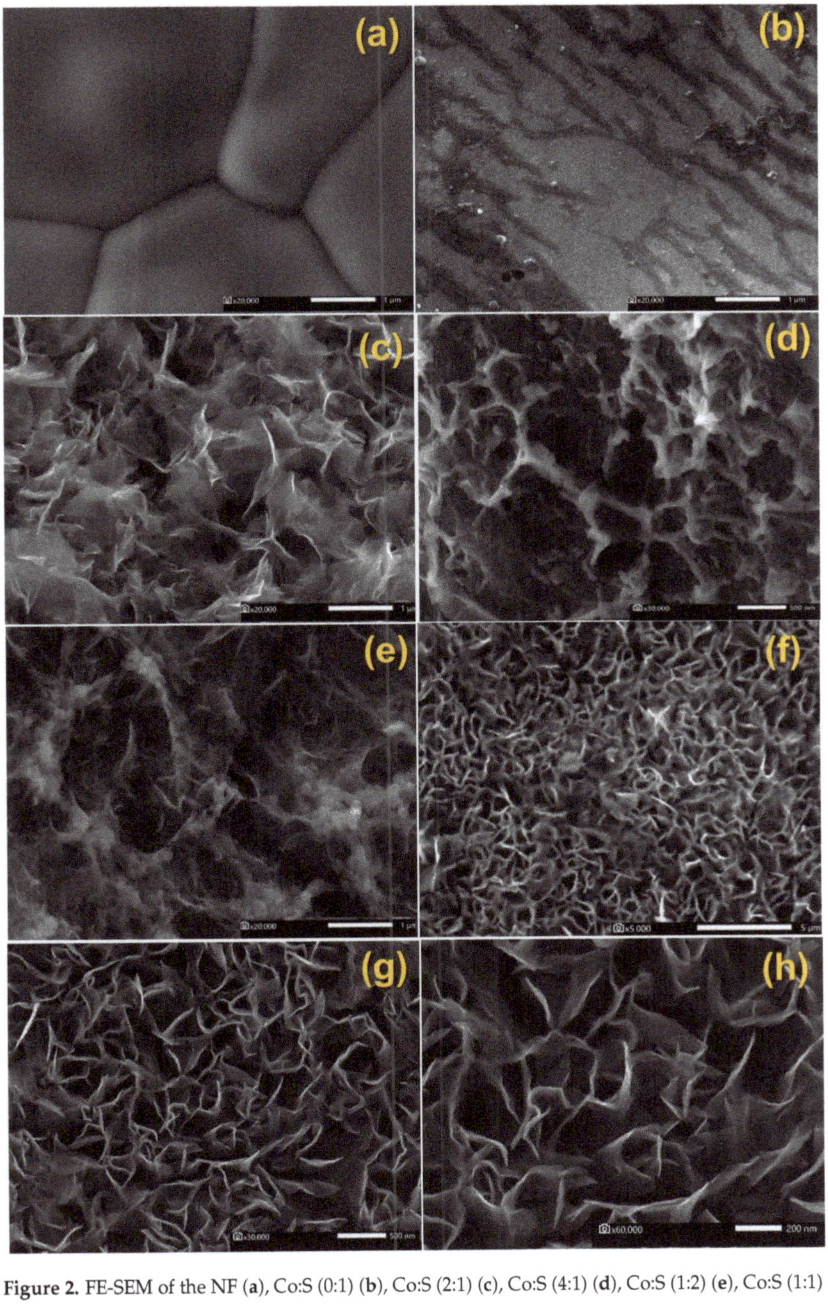

Figure 2. FE-SEM of the NF (**a**), Co:S (0:1) (**b**), Co:S (2:1) (**c**), Co:S (4:1) (**d**), Co:S (1:2) (**e**), Co:S (1:1) (**f**), Co:S (1:1) higher magnification scale (**g**,**h**).

3.3. Electrochemical Evaluation

Over the past years, several mechanisms for the OER in alkaline water electrolysis cells have been proposed [38,39]. Generally, the OER proceeds via two different routes: (a) nucleophilic water attack (WNA) or the adsorbate evolution mechanism (AEM), where MOOH (M is the active site) is initially created, and the reaction proceeds via adsorption/desorption, (b) O–O coupling, followed by O_2 evolution from the coupling of two metal-oxyl radicals with different variants [24,40,41].

The following reaction scheme Is usually considered for the OER (WNA) in an alkaline medium [24]:

$$M + OH^- \rightarrow MOH + e^- \tag{8}$$

$$MOH + OH^- \rightarrow MO + H_2O + e^- \tag{9}$$

$$MO + OH^- \rightarrow MOOH + e^- \text{ (WNA)} \tag{10}$$

$$MOOH + OH^- \rightarrow M + O_2 + H_2O + e^- \text{ (WNA)} \tag{11}$$

Finally, the overall reaction can be written as follows:

$$4OH^- \rightarrow O_2 + 2H_2O + 4e^- \tag{12}$$

Furthermore, recent studies have proposed a new mechanism for the OER based on redox chemistry, defined as a lattice oxidation mechanism (LOM). In this mechanism, the lattice oxygen could be activated at the corresponding potential and could participate in the formation of O–O active intermediates during the OER [42–44].

In addition, when transition metal chalcogenides (TMCs), such as S, are combined with transition metals such as Co, a kinetic structural transformation of the Co_xS_y occurs during the OER [45]. An underlying dissolution of sulfur atoms promotes the oxidation of cobalt ions and facilitates the transformation of Co_xS_y nanovesicles to $Co(OH)_2$ and then to crystalline CoOOH, initiating the OER [46,47].

In order to explore the OER kinetics of as-prepared electrocatalysts (Figure 3a), the electrochemical polarization curves were obtained in 1 M KOH, with a scan rate of 1 mV s^{-1} (90% iR compensation) (Figure 3a). The obtained results, summarized in Table 2, clearly show the significant impact of the Co:S ratio on the OER kinetics. When only S is deposited on NF (Co:S (0:1) sample), a significant inhibition is observed, reflected in overpotential. On the other hand, the deposition of bare cobalt over the NF substrate (Co:S (1:0) sample) has a positive effect, resulting in a lower overpotential (Figure 3a, Table 2). Notably, the fabricated electrodes containing both Co and S demonstrate lower overpotential values, implying the synergistic effect of Co_xS_y@NF composites for an enhanced OER [48]. It is evident that the Co:S (1:1) electrode, with an equimolar Co and S content, exhibits the lowest overpotential value of 0.28 V at a 10 mA cm^{-2} current density, $|\eta_{10}|$, being far lower compared to 0.42 V of bare NF.

Table 2. Overpotential values ($|\eta_{10}|$) and Tafel slope in 1M KOH of all the fabricated electrodes.

Electrode	$\|\eta_{10}\|$ (V)	Tafel Slope (mV dec^{-1})
NF	0.42	134 ± 0.01
Co:S (1:0)	0.31	106 ± 0.7
Co:S (0:1)	0.53	135 ± 0.7
Co:S (1:2)	0.32	113 ± 0.6
Co:S (2:1)	0.34	110 ± 0.4
Co:S (4:1)	0.3	127 ± 0.7
Co:S (1:1)	0.28	95 ± 0.3

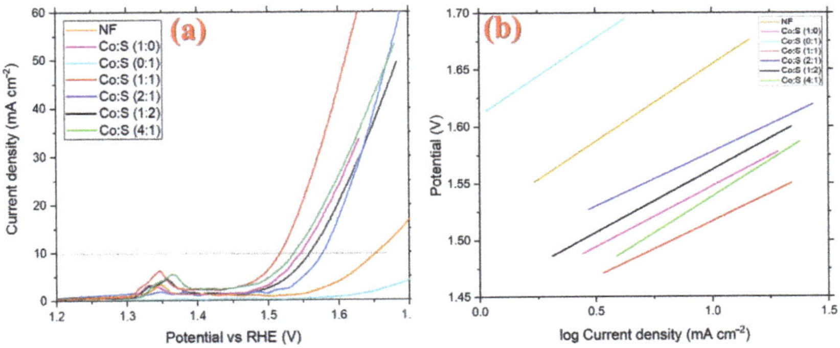

Figure 3. Electrochemical performance of the NF, Co:S (1:0), Co:S (0:1), Co:S (1:1), Co:S (2:1), Co:S (1:2), Co:S (4:1) for the OER: (**a**) Linear sweep voltammograms (LSV) at a scan rate of 1 mV s^{-1} in 1 M KOH; (**b**) Tafel plots of the fabricated electrodes corresponding to the LSV curves.

At this point, it is also worth noting that the S−doped NF electrodes (Co:S (0:1)), where NiS is mainly detected, exhibit an even inferior performance compared to bare NF, plausibly implying the low OER reactivity of the NiS phase under the present conditions. On the other hand, the Co:S (1:0) electrode, where Co_xS_y has been detected, exhibits an adequate OER performance, most probably revealing the high reactivity of the Co_xS_y phase towards the OER reaction. Although these findings cannot exclude the possibility of the synergistic effects between the Ni_xS_x and Co_xS_y phases when both co-exist on the NF surface, they imply the pivotal effect of Co and S coexistence on the improvement in OER performance.

Afterward, the Tafel slope was determined using the linear part of the Tafel plot, according to the following equation:

$$\eta = a + b \log j \tag{13}$$

where n is the overpotential value, j is the current density, a is the fitting parameter, and b is the Tafel slope [4,16,49].

The calculated Tafel slopes of the fabricated electrodes (Figure 3b and Figure S2 and Table 2) further strengthen the findings regarding the enhanced electrocatalytic activity of the Co:S (1:1) electrode, presenting the lowest Tafel slope (95 mV dec^{-1}) for the OER. The lowest Tafel slope of the Co:S (1:1) electrode confirms its faster oxygen evolution reaction kinetics, which could be attributed to the electrode's distinct nano-architecture (Figure 2f), offering the required in-plane pores for a better electrolyte flow and an extension of the active electrocatalytic area for the OER. Moreover, the Co:S (1:1) shows excellent stability in a 30 wt% KOH solution for more than 12 h (Figure S4), demonstrating its potential for practical applications. In addition, the lower overpotential of the Co:S (1:1) at 50 mA cm^{-2} compared to the other Co_xS_y electrodes is obvious (Figure S4); this is due to the retainment of flower-like nanosheets on its surface after a prolonged stability test (Figure S5). Also, the almost unchanged XRF spectra of the Co:S (1:1) electrode, before and after the stability test, clearly reveal the stability of the aforementioned electrode (Figure S5). These findings demonstrate the stability of CoxSy-based electrodes in terms of their morphological and compositional characteristics, leading to an enhanced OER performance.

The superior performance of the Co:S (1:1) electrode can be further revealed through a comparison with state-of-the-art Co_xS_y-containing electrodes over NF substrates, as well as with the benchmark Ru and Ru@IrO_2 electrodes used for the OER (Table 3). It is evident that the as-prepared Co_xS_y@NF electrodes exhibit lower values compared to the Ru and Ru@IrO_2 catalysts, but are superior compared to most of the non-noble metal catalysts, offering one of the lowest Tafel slopes despite their simpler composition (monometallic Co electrodes) and facile fabrication procedure (one-step electrodeposition).

Table 3. Comparison with state-of-the-art Co- and S-containing electrodes for the OER in alkaline medium.

Electrodes	Method	Tafel Slope mV dec^{-1}	Electrolyte	Reference
$NiCo_2S_4/Ni_3S_2$	Hydrothermal	137	1 M KOH	[50]
NiCo-LDHs	Hydrothermal	118	1 M KOH	[51]
$Co\text{-}Ni_3S_2/NF$	Hydrothermal method–liquid-phase vulcanization	176	1 M KOH	[52]
$CoNiS_x/NF$	Sulfuration process	107	1 M KOH	[53]
$NiCo_2S_4/NF$	Hydrothermal	95	1 M KOH	[54]
Co_9S_8 NM/NF	Hydrothermal	150	1 M KOH	[55]
$NiCo_2S_4$-NF	Hydrothermal	91	1 M NaOH	[56]
Ru nanoparticles	Laser-generated	70	0.5 M H_2SO_4	[57]
$Ru@IrO_x$	Charge redistribution	69	0.05 M H_2SO_4	[58]
Co:S (1:1)@NF	Electrodeposition	95	1 M KOH	This work

The enhanced OER kinetics of the Co:S (1:1) electrodes can be ascribed to the formation of flower-like nanosheets over the NF substrate (Figure 2e,f); these increase the electrochemically active area and are responsible for the high utilization ratio of electrocatalytic active sites [16]. Moreover, the absence of a binder and the direct adherence of Co and S onto the NF substrate may further account for the facile electron transportation and improved electrical conductivity, as supported by EIS experiments (see below).

Moreover, to verify our assumption that the mean particle size is closely linked to the Co/S molar ratio and, in turn, to the electrochemical performance, the Tafel slope and mean grain size are plotted in Figure 4 as a function of the Co/S ratio. Notably, the mean particle size totally coincides with the Tafel slope, implying their interrelation; the small particle size offers more electrochemical active sites, facilitating the OER kinetics. Neither Co-rich nor S-rich electrodes provide the optimum performance, maximized for the equimolar Co:S ratio of 1:1; this is in perfect agreement with the electrochemical performance results (lowest Tafel slope and overpotential values).

The double-layer capacitance (C_{DL}) and electrochemical active surface area (ECSA) were next considered in order to gain insights into the intrinsic electrocatalytic reactivity of the as-prepared electrodes. The C_{DL}, obtained via the cyclic voltammetry method, is proportional to the dependent capacitive current (J_{DL}) in the following equation:

$$J_{DL} = C_{DL} \times \frac{v}{A} \tag{14}$$

where v stands for the scan rate (V s^{-1}) and A for the electrode surface (cm^2) [39,59,60].

The following equation can be used to obtain the ESCA values:

$$ESCA = C_{DL} \times \frac{1}{20} \tag{15}$$

The value 20 μF cm^{-2} is the C_{DL} value of a perfectly smooth Ni electrode [61,62].

Cyclic voltammograms with scan rates of 5 to 100 mVs^{-1} (Figure S4) were obtained close to the open circuit potential (±50 mV) of each electrode in a non-Faradaic region to calculate the C_{DL} values [63–65] (Figure 5, Table 4). Afterward, the C_{DL} and the ESCA values were calculated by plotting the average J_{DL} vs. scan rate and obtaining the slope.

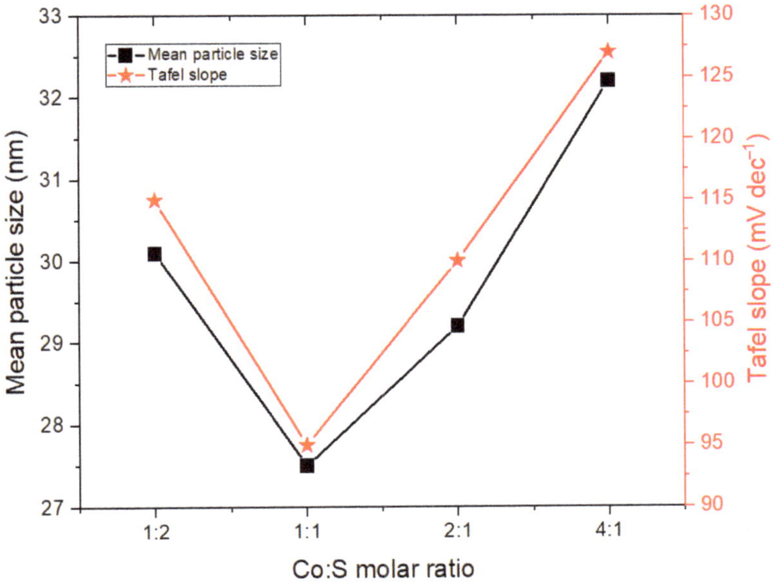

Figure 4. Co:S molar ratio vs. mean particle size and Tafel slope of the Co:S (1:1), Co:S (2:1), Co:S (1:2), Co:S (4:1) electrodes.

Table 4. Double-layer capacitance and ECSA values for the as-synthesized electrodes.

Electrode	C_{DL} Value (mF cm^{-2})	ECSA (cm^2)
NF	0.76 ± 0.13	38 ± 0.5
Co:S (1:0)	1.96 ± 0.13	98 ± 0.4
Co:S (0:1)	0.26 ± 0.01	13 ± 0.3
Co:S (1:2)	1.67 ± 0.09	83.5 ± 0.5
Co:S (2:1)	2.16 ± 0.23	108 ± 0.8
Co:S (4:1)	1.78 ± 0.19	89 ± 0.4
Co:S (1:1)	10.74 ± 0.71	537 ± 1.1

Cobalt deposition generally increases the C_{DL} and ESCA values compared to the NF reference electrode. On the other hand, sulfur deposition has a detrimental effect. Remarkably, the Co:S (1:1) electrode exhibits the best C_{DL} (10.74 mF cm^{-2}) and ESCA (537 cm^2) values, which are about one order of magnitude higher compared to the corresponding values of the NF background electrode. These findings are in line with the optimum structural (smallest particle size) and morphological (highly dispersed 3D nanosheets) characteristics of the Co:S (1:1) electrode, further corroborating the close relationship between the structural–morphological features and electrochemical performance. It is also worth noting that the high C_{DL} values of the Co_xS_y@NF electrodes imply their potential use as electrodes for electrochemical supercapacitors [66,67].

Electrochemical impedance spectroscopy (EIS) measurements were next carried out to gain insights into the charge transfer and transport processes involved in the OER reaction. EIS was applied in potentiostatic mode between frequencies ranging from 10 kHz to 0.1 Hz, applying a sinusoidal alternating current (AC) potential of 10 mV (RMS). In Figure 6, the Nyquist plots of the as-prepared electrodes are presented.

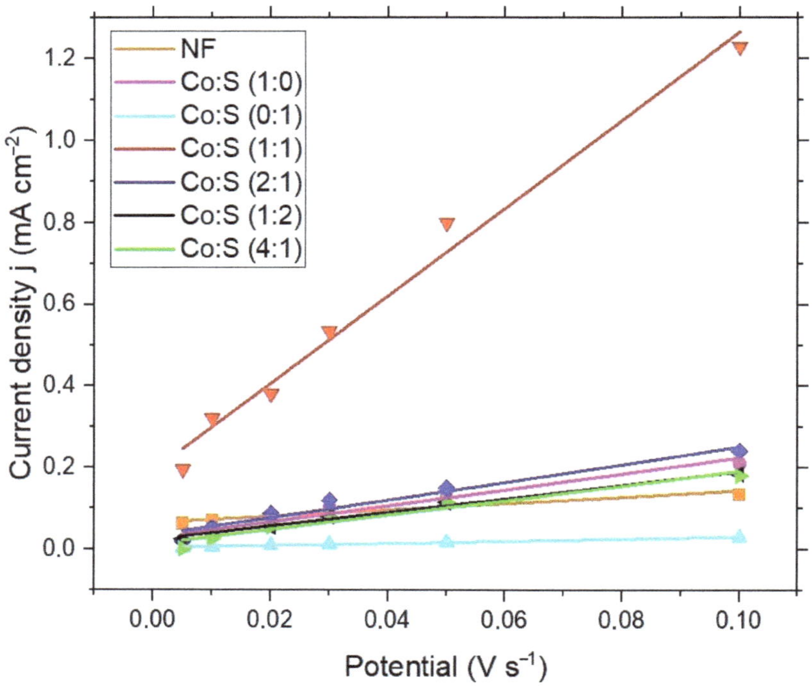

Figure 5. Average absolute current density vs. scan rate for C_{DL} estimation of the NF, Co:S (1:0), Co:S (0:1), Co:S (1:1), Co:S (2:1), Co:S (1:2), and Co:S (4:1) electrodes, at scan rates of 0.005 V s^{-1}–0.1 V s^{-1}.

The intercept point on the Z real axis of the Nyquist plot (Figure 6), at high frequencies, depicts the electrolyte and the internal electrode's resistance (R_s) [68–70]. The gradient line achieved in low frequencies corresponds to the diffusion resistance known as the Warburg element, W [71]. The constant phase element (CPE) represents the double-layer capacitance between solid and ionic solutions [72]. Also, the charge transfer resistance (R_{ct}) reflects the difficulty the charge transfer oxygen evolution reaction has in proceeding. Table 5 summarizes the R_s and R_{ct} values of all fabricated electrodes.

Table 5. Charge transfer resistance (R_{ct}) and electronic resistance R_s of Co:S (1:0), Co:S (0:1), Co:S (1:2), Co:S (2:1), Co:S (4:1), and Co:S (1:1) electrodes.

Electrode	R_s (Ω)	R_{ct} (Ω)
Co:S (1:0)	2.1 ± 0.003	39.4 ± 9.5
Co:S (0:1)	2.8 ± 0.004	155,566 ± 3449
Co:S (1:2)	2.5 ± 0.004	1.9 ± 0.2
Co:S (2:1)	2 ± 0.009	521 ± 7
Co:S (4:1)	2.3 ± 0.01	0.75 ± 0.03
Co:S (1:1)	2.2 ± 0.008	0.000006 ± 0.008

It is apparent that the Co:S (1:1) electrode provides, by far, the lower R_{ct}, resulting in an enhancement in the OER kinetics. Moreover, at higher frequencies, the Co:S (1:1) electrode exhibits the Warburg line (Figure 6), implying a better capacity and improved conductivity [71,73]. It should be noted, however, that no significant differences can be observed between the R_s values of Co-containing samples, implying the crucial role of the active metal phase (Co) in determining the electrode's resistance. Hence, based on the present results, the excellent electrochemical performance of the Co:S (1:1) electrode could be mainly ascribed to its low R_{ct}, thereby accelerating the charge transfer oxygen evolution reaction.

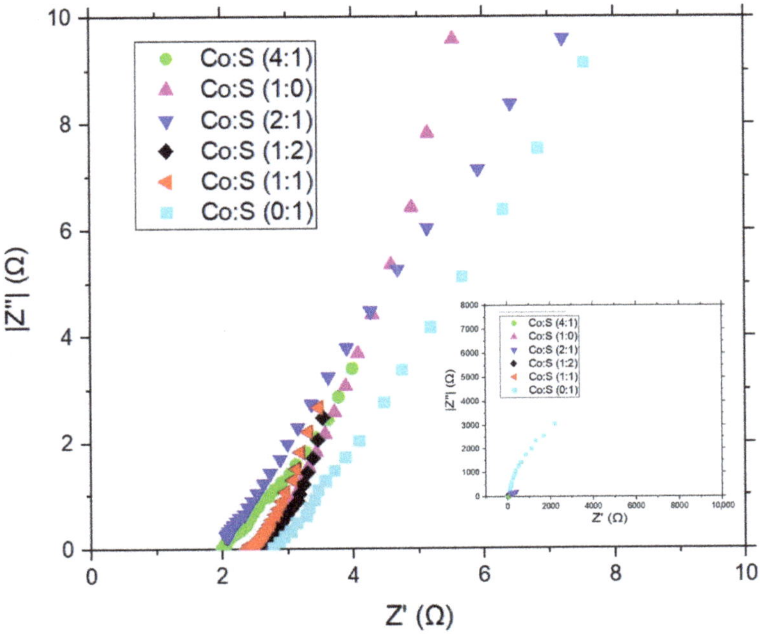

Figure 6. Zoomed in Nyquist plot of Co:S (1:0), Co:S (0:1), Co:S (1:2), Co:S (2:1), Co:S (4:1), and Co:S (1:1) electrodes in 1M KOH. Insets: zoomed out Nyquist plot.

4. Conclusions

The present work systematically explored the impact of the Co/S molar ratio on the OER kinetics and electrochemical performance of Co_xS_y@NF electrodes. Thiourea and cobalt nitrate precursors at different stoichiometric concentrations were used to fabricate Co_xS_y@NF heterostructures through the electrodeposition method.

Notably, it was found that the Co:S ratio profoundly influenced the structure and morphology of the as-prepared electrodes and, in turn, their OER kinetics. The Co_xS_y@NF electrode with a Co:S ratio of 1:1 exhibited the lowest overpotential value at 10 mA cm^{-2} (0.28 V) and a Tafel slope of 95 mV dec^{-1}, offering, in addition, a high double-layer capacitance (C_{DL}) of 10.74 mF cm^{-2}. Electrochemical impedance spectroscopy confirmed the key role of the Co:S ratio on charge transfer resistance, which is substantially decreased at a Co:S molar ratio of 1:1. Structural and morphological analysis disclosed that the Co_xS_y@NF electrocatalyst with an equimolar Co:S ratio presented a 3D flower-like nanosheet morphology, offering the smallest particle size and the highest electrochemical active area, which are both conducive to improving the OER kinetics.

The enhanced electrocatalytic activity, stability, facile synthesis route, and free-of-binders fabrication procedure render Co_xS_y@NF electrodes a promising earth-abundant, noble metal-free, bifunctional type of electrocatalyst for the OER in alkaline water electrolysis cells.

Supplementary Materials: The following supporting information can be downloaded at: https://www.mdpi.com/article/10.3390/surfaces6040033/s1, Figure S1. XRF analysis of the Co:S (1:1) electrode; Figure S2. EDS analysis of the as-prepared Co:S electrodes; Figure S3. Tafel slope of the fabricated electrodes in 1 M KOH electrolyte; Figure S4. Stability test of all the fabricated electrodes; Figure S5. FE-SEM of the Co:S (2:1) (a), Co:S (4:1) (b), Co:S (1:2) (c), Co:S (1:1) after prolonged stability tests. XRF spectra of the Co:S (1:1) electrode before and after the OER stability test; Figure S6. Capacitance double layer measurements with cyclic voltammetry; Figure S7. Eis fitting data and equivalent circuits for the as-prepared electrodes; Table S1. Lattice parameters of the as-prepared electrodes.

Author Contributions: Conceptualization, I.P. and M.K.; methodology, I.P. and M.K.; validation, I.P., M.K. and N.P.; formal analysis, I.P., M.K.; investigation, I.P.; resources, M.K., P.A.L. and G.E.M.; data curation, I.P.; writing—original draft preparation, I.P.; writing—review and editing, I.P., M.K., G.E.M.; visualization, I.P. and M.K.; supervision, M.K., G.E.M. and P.A.L.; project administration, M.K.; funding acquisition, M.K., G.E.M. and P.A.L. All authors have read and agreed to the published version of the manuscript.

Funding: This research has received funding from the European Union under grant agreement No 101099717–ECOLEFINS project. Views and opinions expressed are, however, those of the author(s) only and do not necessarily reflect those of the European Union or European Innovation Council and SMEs Executive Agency (EISMEA) granting authority. Neither the European Union nor the granting authority can be held responsible for them. PL, NP acknowledge financial support from the European Union's Horizon 2020 research and innovation program under grant agreement no 871124 Laserlab-Europe. The authors would like to acknowledge the HELLAS-CH national infrastructure (MIS 5002735) implemented under "Action for Strengthening Research and Innovation Infrastructures," funded by the Operational Programme "Competitiveness, Entrepreneurship, and Innovation" (NSRF 2014-2020) and co-financed by Greece and the European Union (European Regional Development Fund).

Institutional Review Board Statement: Not applicable.

Informed Consent Statement: Not applicable.

Data Availability Statement: The data supporting the findings of this study are available in the form of Tables/Figures within the article and its Supplementary Materials. The raw data are available from the corresponding authors upon reasonable request.

Acknowledgments: The authors want to acknowledge A. Manousaki for the field emission SEM.

Conflicts of Interest: The authors declare no conflict of interest.

References

1. Bockris, J.O.; Veziroglu, T.N. Estimates of the price of hydrogen as a medium for wind and solar sources. *Int. J. Hydrogen Energy* **2007**, *32*, 1605–1610. [CrossRef]
2. Guo, Y.; Shang, C.; Li, J.; Wang, E. Recent development of hydrogen evolution, oxygen evolution and oxygen reduction reaction. *Sci. Sin. Chim.* **2018**, *48*, 926–940. [CrossRef]
3. Wei, Y.; Wang, M.; Fu, W.; Wei, L.; Zhao, X.; Zhou, X.; Ni, M.; Wang, H. Highly active and durable catalyst for hydrogen generation by the NaBH4 hydrolysis reaction: CoWB/NF nanodendrite with an acicular array structure. *J. Alloys Compd.* **2020**, *836*, 155429. [CrossRef]
4. Poimenidis, I.A.; Papakosta, N.; Klini, A.; Farsari, M.; Konsolakis, M.; Loukakos, P.A.; Moustaizis, S. Electrodeposited Ni foam electrodes for increased hydrogen production in alkaline electrolysis. *Fuel* **2023**, *342*, 127798. [CrossRef]
5. Guan, D.; Wang, B.; Zhang, J.; Shi, R.; Jiao, K.; Li, L.; Wang, Y.; Xie, B.; Zhang, Q.; Yu, J.; et al. Hydrogen society: From present to future. *Energy Environ. Sci.* **2023**, *16*, 4926–4943. [CrossRef]
6. McCrory, C.C.L.; Jung, S.; Ferrer, I.M.; Chatman, S.M.; Peters, J.C.; Jaramillo, T.F. Benchmarking Hydrogen Evolving Reaction and Oxygen Evolving Reaction Electrocatalysts for Solar Water Splitting Devices. *J. Am. Chem. Soc.* **2015**, *137*, 4347–4357. [CrossRef]
7. Han, G.-Q.; Shang, X.; Lu, S.-S.; Dong, B.; Li, X.; Liu, Y.-R.; Hu, W.; Zeng, J.-B.; Chai, Y.-M.; Liu, C.-G. Electrodeposited MoSx films assisted by liquid crystal template with ultrahigh electrocatalytic activity for hydrogen evolution reaction. *Int. J. Hydrogen Energy* **2017**, *42*, 5132–5138. [CrossRef]
8. Wu, G.; Santandreu, A.; Kellogg, W.; Gupta, S.; Ogoke, O.; Zhang, H.; Wang, H.-L.; Dai, L. Carbon nanocomposite catalysts for oxygen reduction and evolution reactions: From nitrogen doping to transition-metal addition. *Nano Energy* **2016**, *29*, 83–110. [CrossRef]
9. Kanan, M.W.; Nocera, D.G. In Situ Formation of an Oxygen-Evolving Catalyst in Neutral Water Containing Phosphate and Co^{2+}. *Science (1979)* **2008**, *321*, 1072–1075. [CrossRef]
10. Song, J.; Wei, C.; Huang, Z.-F.; Liu, C.; Zeng, L.; Wang, X.; Xu, Z. A review on fundamentals for designing oxygen evolution electrocatalysts. *Chem. Soc. Rev.* **2020**, *49*, 2196–2214. [CrossRef]

11. Imran Anwar, M.; Manzoor, S.; Ma, L.; Asad, M.; Zhang, W.; Shafiq, Z.; Ashiq, M.-N.; Yang, G. Nitrogen-rich three-dimensional metal-organic framework microrods as an efficient electrocatalyst for oxygen evolution reaction and supercapacitor applications. *Fuel* **2023**, *331*, 125746. [CrossRef]
12. Jung, S.; McCrory, C.C.L.; Ferrer, I.M.; Peters, J.C.; Jaramillo, T.F. Benchmarking nanoparticulate metal oxide electrocatalysts for the alkaline water oxidation reaction. *J. Mater. Chem. A Mater.* **2016**, *4*, 3068–3076. [CrossRef]
13. Chi, J.-Q.; Shang, X.; Liang, F.; Dong, B.; Li, X.; Liu, Y.-R.; Yan, K.-L.; Gao, W.-K.; Chai, Y.-M.; Liu, C.-G. Facile synthesis of pyrite-type binary nickel iron diselenides as efficient electrocatalyst for oxygen evolution reaction. *Appl. Surf. Sci.* **2017**, *401*, 17–24. [CrossRef]
14. Sarfraz, B.; Bashir, I.; Rauf, A. CuS/NiFe-LDH/NF as a Bifunctional Electrocatalyst for Hydrogen Evolution (HER) and Oxygen Evolution Reactions (OER). *Fuel* **2023**, *337*, 127253. [CrossRef]
15. Shit, S.; Chhetri, S.; Jang, W.; Murmu, N.C.; Koo, H.; Samanta, P.; Kuila, T. Cobalt Sulfide/Nickel Sulfide Heterostructure Directly Grown on Nickel Foam: An Efficient and Durable Electrocatalyst for Overall Water Splitting Application. *ACS Appl. Mater. Interfaces* **2018**, *10*, 27712–27722. [CrossRef] [PubMed]
16. Kale, S.B.; Lokhande, A.C.; Pujari, R.B.; Lokhande, C.D. Cobalt sulfide thin films for electrocatalytic oxygen evolution reaction and supercapacitor applications. *J. Colloid Interface Sci.* **2018**, *532*, 491–499. [CrossRef]
17. Kuai, L.; Geng, J.; Chen, C.; Kan, E.; Liu, Y.; Wang, Q.; Geng, B. A Reliable Aerosol-Spray-Assisted Approach to Produce and Optimize Amorphous Metal Oxide Catalysts for Electrochemical Water Splitting. *Angew. Chem. Int. Ed.* **2014**, *53*, 7547–7551. [CrossRef]
18. Smith, R.D.L.; Prévot, M.S.; Fagan, R.D.; Zhang, Z.; Sedach, P.A.; Siu, M.K.J.; Trudel, S.; Berlinguette, C. Photochemical Route for Accessing Amorphous Metal Oxide Materials for Water Oxidation Catalysis. *Science* **2013**, *340*, 60–63. [CrossRef]
19. Borthakur, P.; Boruah, P.K.; Das, M.R.; Ibrahim, M.M.; Altalhi, T.; El-Sheshtawy, H.S.; Szunerits, S.; Boukherroub, R.; Amin, M. CoS_2 Nanoparticles Supported on rGO, g-C_3N_4, BCN, MoS_2, and WS_2 Two-Dimensional Nanosheets with Excellent Electrocatalytic Performance for Overall Water Splitting: Electrochemical Studies and DFT Calculations. *ACS Appl. Energy Mater.* **2021**, *4*, 1269–1285. [CrossRef]
20. Poimenidis, I.A.; Lykaki, M.; Moustaizis, S.; Loukakos, P.; Konsolakis, M. One-step solvothermal growth of NiO nanoparticles on nickel foam as a highly efficient electrocatalyst for hydrogen evolution reaction. *Mater. Chem. Phys.* **2023**, *305*, 128007. [CrossRef]
21. Surendran, S.; Shanmugapriya, S.; Lee, Y.S.; Sim, U.; Selvan, R.K. Carbon-Enriched Cobalt Phosphide with Assorted Nanostructure as a Multifunctional Electrode for Energy Conversion and Storage Devices. *ChemistrySelect* **2018**, *3*, 12303–12313. [CrossRef]
22. Prabakaran, K.; Lokanathan, M.; Kakade, B. Three dimensional flower like cobalt sulfide (CoS)/functionalized MWCNT composite catalyst for efficient oxygen evolution reactions. *Appl. Surf. Sci.* **2019**, *466*, 830–836. [CrossRef]
23. Poimenidis, I.A.; Papakosta, N.; Klini, A.; Farsari, M.; Moustaizis, S.D.; Konsolakis, M.; Loukakos, P. Ni foam electrodes decorated with Ni nanoparticles via pulsed laser deposition for efficient hydrogen evolution reaction. *Mater. Sci. Eng. B* **2024**, *299*, 116922. [CrossRef]
24. Zhao, Y.; Adiyeri Saseendran, D.P.; Huang, C.; Triana, C.A.; Marks, W.R.; Chen, H.; Zhao, H.; Patzke, G.-R. Oxygen Evolution/Reduction Reaction Catalysts: From In Situ Monitoring and Reaction Mechanisms to Rational Design. *Chem. Rev.* **2023**, *123*, 6257–6358. [CrossRef]
25. Wang, M.; Zhang, L.; He, Y.; Zhu, H. Recent advances in transition-metal-sulfide-based bifunctional electrocatalysts for overall water splitting. *J. Mater. Chem. A Mater.* **2021**, *9*, 5320–5363. [CrossRef]
26. Zhu, Y.; Song, L.; Song, N.; Li, M.; Wang, C.; Lu, X. Bifunctional and Efficient CoS_2–C@MoS_2 Core–Shell Nanofiber Electrocatalyst for Water Splitting. *ACS Sustain. Chem. Eng.* **2019**, *7*, 2899–2905. [CrossRef]
27. Elshahawy, A.M.; Li, X.; Zhang, H.; Hu, Y.; Ho, K.H.; Guan, C.; Wang, J. Controllable $MnCo_2S_4$ nanostructures for high performance hybrid supercapacitors. *J. Mater. Chem. A Mater.* **2017**, *5*, 7494–7506. [CrossRef]
28. Yu, Y. Study on Electrochemistry and Nucleation Process of Nickel Electrodeposition. *Int. J. Electrochem. Sci.* **2017**, *12*, 485–495. [CrossRef]
29. Shi, J.; Li, X.; He, G.; Zhang, L.; Li, M. Electrodeposition of high-capacitance 3D CoS/graphene nanosheets on nickel foam for high-performance aqueous asymmetric supercapacitors. *J. Mater. Chem. A Mater.* **2015**, *3*, 20619–20626. [CrossRef]
30. Sultana, U.K.; He, T.; Du, A.; O'Mullane, A.P. An amorphous dual action electrocatalyst based on oxygen doped cobalt sulfide for the hydrogen and oxygen evolution reactions. *RSC Adv.* **2017**, *7*, 54995–55004. [CrossRef]
31. Jiang, S.P.; Chen, Y.Z.; You, J.K.; Chen, T.X.; Tseung, A.C.C. Reactive Deposition of Cobalt Electrodes: I. Experimental. *J. Electrochem. Soc.* **1990**, *137*, 3374–3380. [CrossRef]
32. García-Valenzuela, J.A. Simple Thiourea Hydrolysis or Intermediate Complex Mechanism? Taking up the Formation of Metal Sulfides from Metal–Thiourea Alkaline Solutions. *Comments Inorg. Chem.* **2017**, *37*, 99–115. [CrossRef]
33. Zhang, J.; Xu, C.; Zhang, D.; Zhao, J.; Zheng, S.; Su, H.; Wei, F.; Yuan, B.; Fernandez, C. Facile Synthesis of a Nickel Sulfide (NiS) Hierarchical Flower for the Electrochemical Oxidation of H_2O_2 and the Methanol Oxidation Reaction (MOR). *J. Electrochem. Soc.* **2017**, *164*, B92–B96. [CrossRef]
34. Liu, S.; Shi, Q.; Tong, J.; Li, S.; Li, M. Controlled synthesis of spherical α-NiS and urchin-like β-NiS microstructures. *J. Exp. Nanosci.* **2014**, *9*, 475–481. [CrossRef]

35. Yan, K.-L.; Shang, X.; Li, Z.; Dong, B.; Chi, J.-Q.; Liu, Y.-R.; Gao, W.-K.; Chai, Y.-M.; Liu, C.-G. Facile synthesis of binary NiCoS nanorods supported on nickel foam as efficient electrocatalysts for oxygen evolution reaction. *Int. J. Hydrogen Energy* **2017**, *42*, 17129–17135. [CrossRef]
36. Samal, R.; Mondal, S.; Gangan, A.S.; Chakraborty, B.; Rout, C.S. Comparative electrochemical energy storage performance of cobalt sulfide and cobalt oxide nanosheets: Experimental and theoretical insights from density functional theory simulations *Phys. Chem. Chem. Phys.* **2020**, *22*, 7903–7911. [CrossRef] [PubMed]
37. Huang, S.; He, Q.; Liu, M.; An, H.; Hou, L. Controlled synthesis of porous nanosheets-assembled peony-like cobalt nickel selenides for triiodide reduction in dye-sensitized solar cells. *J. Alloys Compd.* **2020**, *818*, 152817. [CrossRef]
38. Shervedani, R.K.; Madram, A.R. Kinetics of hydrogen evolution reaction on nanocrystalline electrodeposited $Ni_{62}Fe_{35}C_3$ cathode in alkaline solution by electrochemical impedance spectroscopy. *Electrochim. Acta* **2007**, *53*, 426–433. [CrossRef]
39. Hosseini, M.G.; Momeni, M.M.; Faraji, M. Highly Active Nickel Nanoparticles Supported on TiO_2 Nanotube Electrodes for Methanol Electrooxidation. *Electroanalysis* **2010**, *22*, 2620–2625. [CrossRef]
40. Shaffer, D.W.; Xie, Y.; Concepcion, J.J. O–O bond formation in ruthenium-catalyzed water oxidation: Single-site nucleophilic attack vs. O–O radical coupling. *Chem. Soc. Rev.* **2017**, *46*, 6170–6193. [CrossRef]
41. Craig, M.J.; Coulter, G.; Dolan, E.; Soriano-López, J.; Mates-Torres, E.; Schmitt, W.; Melchor, M.-G. Universal scaling relations for the rational design of molecular water oxidation catalysts with near-zero overpotential. *Nat. Commun.* **2019**, *10*, 4993. [CrossRef] [PubMed]
42. Zhao, M.; Zheng, X.; Cao, C.; Lu, Q.; Zhang, J.; Wang, H.; Huang, Z.; Cao, Y.; Wang, Y.; Deng, Y. Lattice oxygen activation in disordered rocksalts for boosting oxygen evolution. *Phys. Chem. Chem. Phys.* **2023**, *25*, 4113–4120. [CrossRef]
43. Zhu, Y.; Tahini, H.A.; Hu, Z.; Chen, Z.; Zhou, W.; Komarek, A.C.; Lin, Q.; Lin, H.-J.; Chen, C.-T.; Zhong, Y.; et al. Boosting Oxygen Evolution Reaction by Creating Both Metal Ion and Lattice-Oxygen Active Sites in a Complex Oxide. *Adv. Mater.* **2020**, *32*, 1905025. [CrossRef]
44. Yoo, J.S.; Rong, X.; Liu, Y.; Kolpak, A.M. Role of Lattice Oxygen Participation in Understanding Trends in the Oxygen Evolution Reaction on Perovskites. *ACS Catal.* **2018**, *8*, 4628–4636. [CrossRef]
45. Cao, D.; Liu, D.; Chen, S.; Moses, O.A.; Chen, X.; Xu, W.; Wu, C.; Zheng, L.; Chu, S.; Jiang, H.; et al. *Operando* X-ray spectroscopy visualizing the chameleon-like structural reconstruction on an oxygen evolution electrocatalyst. *Energy Environ. Sci.* **2021**, *14*, 906–915. [CrossRef]
46. Fan, K.; Zou, H.; Lu, Y.; Chen, H.; Li, F.; Liu, J.; Sun, L.; Tong, L.; Toney, M.; Sui, M.; et al. Direct Observation of Structural Evolution of Metal Chalcogenide in Electrocatalytic Water Oxidation. *ACS Nano* **2018**, *12*, 12369–12379. [CrossRef] [PubMed]
47. Wang, M.; Dong, C.-L.; Huang, Y.-C.; Shen, S. Operando Spectral and Electrochemical Investigation into the Heterophase Stimulated Active Species Transformation in Transition-Metal Sulfides for Efficient Electrocatalytic Oxygen Evolution. *ACS Catal.* **2020**, *10*, 1855–1864. [CrossRef]
48. Li, M.; Xu, Z.; Li, Y.; Wang, J.; Zhong, Q. In situ fabrication of cobalt/nickel sulfides nanohybrid based on various sulfur sources as highly efficient bifunctional electrocatalysts for overall water splitting. *Nano Sel.* **2022**, *3*, 147–156. [CrossRef]
49. Poimenidis, I.A.; Papakosta, N.; Manousaki, A.; Klini, A.; Farsari, M.; Moustaizis, S.D.; Loukakos, P. Electrodeposited laser–nanostructured electrodes for increased hydrogen production. *Int. J. Hydrogen Energy* **2022**, *47*, 9527–9536. [CrossRef]
50. Wang, Y.; Chen, N.; Du, X.; Han, X.; Zhang, X. Transition metal atoms M (M = Mn, Fe, Cu, Zn) doped nickel-cobalt sulfides on the Ni foam for efficient oxygen evolution reaction and urea oxidation reaction. *J. Alloys Compd.* **2022**, *893*, 162269. [CrossRef]
51. Liu, B.; Zhang, M.; Wang, Y.; Chen, Z.; Yan, K. Facile synthesis of defect-rich ultrathin NiCo-LDHs, NiMn-LDHs and NiCoMn-LDHs nanosheets on Ni foam for enhanced oxygen evolution reaction performance. *J. Alloys Compd.* **2021**, *852*, 156949. [CrossRef]
52. Qin, J.-F.; Yang, M.; Hou, S.; Dong, B.; Chen, T.-S.; Ma, X.; Xie, J.-X.; Zhou, Y.-N.; Nan, J.; Chai, Y.-M. Copper and cobalt co-doped Ni_3S_2 grown on nickel foam for highly efficient oxygen evolution reaction. *Appl. Surf. Sci.* **2020**, *502*, 144172. [CrossRef]
53. Zhang, R.-L.; Duan, J.-J.; Feng, J.-J.; Mei, L.-P.; Zhang, Q.-L.; Wang, A.-J. Walnut kernel-like iron-cobalt-nickel sulfide nanosheets directly grown on nickel foam: A binder-free electrocatalyst for high-efficiency oxygen evolution reaction. *J. Colloid Interface Sci.* **2021**, *587*, 141–149. [CrossRef] [PubMed]
54. Min, K.; Yoo, R.; Kim, S.; Kim, H.; Shim, S.E.; Lim, D.; Baeck, S.-H. Facile synthesis of P-doped $NiCo_2S_4$ nanoneedles supported on Ni foam as highly efficient electrocatalysts for alkaline oxygen evolution reaction. *Electrochim. Acta* **2021**, *396*, 139236. [CrossRef]
55. Gao, W.-K.; Qin, J.-F.; Wang, K.; Yan, K.-L.; Liu, Z.-Z.; Lin, J.-H.; Chai, Y.-M.; Liu, C.-G.; Dong, B. Facile synthesis of Fe-doped Co_9S_8 nano-microspheres grown on nickel foam for efficient oxygen evolution reaction. *Appl. Surf. Sci.* **2018**, *454*, 46–53. [CrossRef]
56. Yin, X.; Sun, G.; Wang, L.; Bai, L.; Su, L.; Wang, Y.; Du, Q.; Shao, G. 3D hierarchical network $NiCo_2S_4$ nanoflakes grown on Ni foam as efficient bifunctional electrocatalysts for both hydrogen and oxygen evolution reaction in alkaline solution. *Int. J. Hydrogen Energy* **2017**, *42*, 25267–25276. [CrossRef]
57. Wang, J.-Q.; Xi, C.; Wang, M.; Shang, L.; Mao, J.; Dong, C.-K.; Liu, H.; Kulinich, S.-A.; Du, X.-W. Laser-Generated Grain Boundaries in Ruthenium Nanoparticles for Boosting Oxygen Evolution Reaction. *ACS Catal.* **2020**, *10*, 12575–12581. [CrossRef]
58. Shan, J.; Guo, C.; Zhu, Y.; Chen, S.; Song, L.; Jaroniec, M.; Zheng, Y.; Qiao, S.-Z. Charge-Redistribution-Enhanced Nanocrystalline Ru@IrOx Electrocatalysts for Oxygen Evolution in Acidic Media. *Chem* **2019**, *5*, 445–459. [CrossRef]
59. Burke, L.D.; Twomey, T.A.M. Voltammetric behaviour of nickel in base with particular reference to thick oxide growth. *J. Electroanal. Chem. Interfacial Electrochem.* **1984**, *162*, 101–119. [CrossRef]

0. Chandrasekaran, P.; Edison, T.N.J.I.; Sethuraman, M.G. Electrocatalytic study of carbon dots/Nickel iron layered double hydroxide composite for oxygen evolution reaction in alkaline medium. *Fuel* **2022**, *320*, 123947. [CrossRef]
1. Navarro-Flores, E.; Chong, Z.; Omanovic, S. Characterization of Ni, NiMo, NiW and NiFe electroactive coatings as electrocatalysts for hydrogen evolution in an acidic medium. *J. Mol. Catal. A Chem.* **2005**, *226*, 179–197. [CrossRef]
2. Li, X.; Liu, P.F.; Zhang, L.; Zu, M.Y.; Yang, Y.X.; Yang, H.G. Enhancing alkaline hydrogen evolution reaction activity through Ni-Mn3O4 nanocomposites. *Chem. Commun.* **2016**, *52*, 10566–10569. [CrossRef]
3. McCrory, C.C.L.; Jung, S.; Peters, J.C.; Jaramillo, T.F. Benchmarking heterogeneous electrocatalysts for the oxygen evolution reaction. *J. Am. Chem. Soc.* **2013**, *135*, 16977–16987. [CrossRef]
4. Poimenidis, I.A.; Tsanakas, M.D.; Papakosta, N.; Klini, A.; Farsari, M.; Moustaizis, S.D.; Loukakos, P.A. Enhanced hydrogen production through alkaline electrolysis using laser-nanostructured nickel electrodes. *Int. J. Hydrogen Energy* **2021**, *46*, 37162–37173. [CrossRef]
5. Symes, D.; Taylor-Cox, C.; Holyfield, L.; Al-Duri, B.; Dhir, A. Feasibility of an oxygen-getter with nickel electrodes in alkaline electrolysers. *Mater. Renew. Sustain. Energy* **2014**, *3*, 27. [CrossRef]
6. Edison, T.N.J.I.; Atchudan, R.; Karthik, N.; Ganesh, K.; Xiong, D.; Lee, Y.R. A novel binder-free electro-synthesis of hierarchical nickel sulfide nanostructures on nickel foam as a battery-type electrode for hybrid-capacitors. *Fuel* **2020**, *276*, 118077. [CrossRef]
7. Huang, H.; Deng, X.; Yan, L.; Wei, G.; Zhou, W.; Liang, X.; Guo, J. One-Step Synthesis of Self-Supported Ni_3S_2/NiS Composite Film on Ni Foam by Electrodeposition for High-Performance Supercapacitors. *Nanomaterials* **2019**, *9*, 1718. [CrossRef]
8. Chen, J.; Wang, Z.; Mu, J.; Ai, B.; Zhang, T.; Ge, W.; Zhang, L. Enhanced lithium storage capability enabled by metal nickel dotted NiO–graphene composites. *J. Mater. Sci.* **2019**, *54*, 1475–1487. [CrossRef]
9. Siddiqui, S.-E.-T.; Rahman, M.d.A.; Kim, J.-H.; Sharif, S.B.; Paul, S. A Review on Recent Advancements of Ni-NiO Nanocomposite as an Anode for High-Performance Lithium-Ion Battery. *Nanomaterials* **2022**, *12*, 2930. [CrossRef]
0. Chia, X.; Ambrosi, A.; Sofer, Z.; Luxa, J.; Pumera, M. Catalytic and Charge Transfer Properties of Transition Metal Dichalcogenides Arising from Electrochemical Pretreatment. *ACS Nano* **2015**, *9*, 5164–5179. [CrossRef]
1. Agudosi, E.S.; Abdullah, E.C.; Numan, A.; Mubarak, N.M.; Aid, S.R.; Benages-Vilau, R.; Romero, P.-G.; Khalid, M.; Omar, N. Fabrication of 3D binder-free graphene NiO electrode for highly stable supercapattery. *Sci. Rep.* **2020**, *10*, 11214. [CrossRef] [PubMed]
2. Conway, B.E.; Birss, V.; Wojtowicz, J. The role and utilization of pseudocapacitance for energy storage by supercapacitors. *J. Power Source* **1997**, *66*, 1–14. [CrossRef]
3. Ge, J.; Wu, J.; Fan, L.; Bao, Q.; Dong, J.; Jia, J.; Guo, Y.; Lin, J. Hydrothermal synthesis of $CoMoO_4/Co_{1-x}S$ hybrid on Ni foam for high-performance supercapacitors. *J. Energy Chem.* **2018**, *27*, 478–485. [CrossRef]

surfaces

MDPI

Communication

Synthesis and Properties of Novel Acrylic Fluorinated Surfactants

Chao Lin [1,*], Jinhua Li [2], Yejun Qin [2], Ping Xing [2] and Biao Jiang [2,*]

1 Shanghai Fire Research Institute of MEM, Shanghai 200032, China
2 Shanghai Institute of Organic Chemistry, Chinese Academy of Sciences, Shanghai 200032, China; lijinhua@mail.sioc.ac.cn (J.L.); qinyj@mail.sioc.ac.cn (Y.Q.); xingping@mail.sioc.ac.cn (P.X.)
* Correspondence: linchao16@163.com (C.L.); jiangb@mail.sioc.ac.cn (B.J.)

Abstract: Branched fluorinated surfactants with creatively introduced acrylate in the hydrophilic group were designed and prepared by adopting perfluoro-2-methyl-2-pentene as the raw substrate. These new compounds showed excellent surface properties, and the surface tension of their aqueous solution at 25 °C could be below 20.00 mN/m at the critical micelle concentration. Compared with similar structures we have synthesized previously, these synthesized compounds exhibit a great improvement with regard to their molecular arrangement at the gas–liquid interface, their polymerizability, and the antibacterial properties of their polymer form, which can provide new ideas in the work to replace perfluorooctane sulfonate/perfluorooctanoic acid.

Keywords: branched fluorinated surfactants; novel acrylic hydrophilic group; surface properties; polymerizable emulsifier; emulsion polymerization

1. Introduction

Fluorinated surfactants were first developed by 3M in the US in the 1950s through electrolytic fluorination under the trade name "Fluorad" [1]. Their unique properties, e.g., their high surface activity and high thermal and chemical stability at low concentrations compare to non-fluorinated ones [2,3], means they have been widely applied in industrial processes and for consumer uses [4–7]. The most common commercially produced fluorinated surfactants are perfluorooctane sulfonate (PFOS)/perfluorooctanoic acid (PFOA) and their derivatives. Their widespread use, disposal, and high stability (they cannot be broken down readily either abiotically or biotically in the environment) have resulted in the widespread presence of per- and polyfluoroalkyl substances (PFASs) in the environment. Commercial productions have shifted toward short-chain alternatives and new fluorinated moieties such as per- and polyfluorinated ethers because perfluoroalkyl substances with chain lengths of C8 or longer have shown great potential to be persistent, toxic, and bioaccumulable [8,9].

Perflur-C6-based fluorine-containing additives do not contain perflur-C8 components and have not been classified as persistent organic pollutants (POPs) previously, but their environmental safety is still controversial. Compared with long-chain analogs, short-chain substitutes with higher water solubility, higher saturated vapor pressure, and lower adsorbability are more likely to migrate in the environment [10,11]. Perfluorohexane sulfonic acid (PFHS) and its related compounds were submitted to the POPs review committee as POP candidates in 2017 [12], and they are now listed in Annex A of the Stockholm Convention without specific exemptions. Substances that potentially degrade to PFHS are also considered as their related compounds [13].

Although other chemical structural compounds with straight-chain perfluorinated hexyl have not been listed in Annex A of the Stockholm Convention, they are the subject of great concern currently. Finding more suitable and reliable compounds and effective preparation methods is very urgent. Our group has carried out numerous studies on preparation processes, structure–activity relationships, and applications using hexafluoropropylene dimer as the raw material [14–16]. The surface tension of this kind of branched fluorinated surfactant at the critical micelle concentration (CMC) is usually around 19 mN/m–20 mN/m, while some compounds with excellent surface activity can reduce the surface tension to below 18 mN/m [16].

To the best of our knowledge, most studies are still focused on reducing the toxicity of fluorinate hydrophobic fragments without paying enough attention to the innovative design of hydrophilic groups. For example, the hydrophilic groups of cationic, amphoteric, and nonionic surfactants are usually quaternary ammonium salt, betaine, and poly-oligomeric ethoxylated alcohol, respectively [9,17]. Designing and synthesizing suitable hydrophilic structures for specific hydrophobic groups to achieve better structural matching are very important tasks in the future for research into PFOS/PFOA alternatives.

2. Materials and Methods

In view of the facts mentioned above, this study proposed a novel hydrophilic group to matching the branched hydrophobic fluorinated tails (Scheme 1). The traditional hydrophilic group structure of quaternary ammonium salt was modified to contain an acrylate structure using an ionic bond or a chemical bond. The existing acrylate structure connected via an ionic bond or chemical bond causes the branched fluorinated surfactants to display significantly different surface activities, thus demonstrating greater application prospects than previous works [14–16], such as more stable foams in fire-fighting agents, polymerizable emulsifiers in emulsion polymerization, and antibacterial additives in coatings.

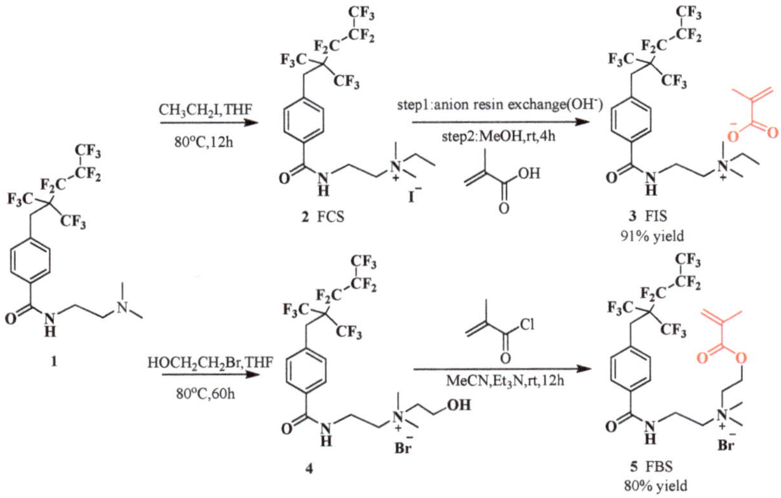

Scheme 1. Synthetic route for the branched fluorosurfactants.

Compound **1** was prepared via four steps according to the literature [15] and was then changed into the ionic paired structural compound **3** (FIS) and the chemical bond structural compound **5** (FBS) via two steps, respectively. For compound **3**, first, quaternary ammonium salt compound **2** underwent quaternarization with iodoethane [15]. Then, conversion of the quaternary ammonium salt **2** into the ionic paired compound **3** was carried out via ion exchange and esterification in turn. For compound **5**, first, quaternary ammonium salt compound **4** underwent quaternarization with bromoethanol. Then, esterification of the alcohol **4** led to the chemical bond structural compound **5**. All of the chemicals and instruments used in this work, the experimental details, and the key spectra are presented in the Supplementary Materials.

3. Results and Discussion

3.1. Surface Activities

The surface or interface tension of surfactants in individual form was tested via the Wilhelmy plate method using a Kruss K100 tensiometer at 25 °C. All values were the average of three measurements. The change trend of surface tension for FIS and FBS in aqueous solutions with various concentrations is presented in Figure 1. The CMC of FIS is 6.57×10^{-4} mol/L and the surface tension at the CMC (γ_{CMC}) is 23.24 mN/m; for FBS, these values are 1.15×10^{-3} mol/L and 19.02 mN/m, respectively. The values of the surface properties of FIS and FBS are better than that of sodium perfluorooctanoate (about 24.7 mN/m at the CMC of 3.1×10^{-2} mol/L) [18].

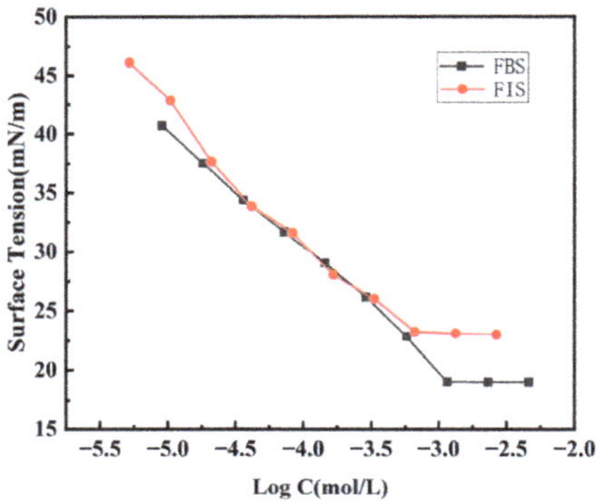

Figure 1. Surface tension measurements of FBS and FIS at 25 °C.

The ability to reduce surface tension is slightly different between FBS and FIS, which can be explained by calculating the surface occupied area per molecule (A_{CMC}). A_{CMC} was derived from the static surface tension vs. $\log_C$ curves through the following equation [19,20]:

$$A_{CMC} = \frac{1}{N_A \Gamma_{\max}} \tag{1}$$

where N_A is Avogadro's number and $\Gamma_{\max}$ is the surface excess concentration as defined by

$$\Gamma_{\max} = -\frac{1}{2.303RT} \log_{c \to c_{CMC}} \left(\frac{d\gamma}{d\lg c} \right) \tag{2}$$

where R is a gas constant and T is the absolute temperature.

The calculated A_{CMC} values of FBS and FIS are listed in Table 1.

Table 1. Static surface and interface properties of different water solutions at 25 °C.

Entry	Systems	CMC (mol/L)	γ_{CMC} (mN/m)	A_{CMC} (Å^2/mol)
1	FIS	6.57×10^{-4}	23.24	2.46
2	FBS	1.15×10^{-3}	19.02	1.65
3 [a]	FCS	2.50×10^{-3}	19.68	52
4 [a]	FAS	1.04×10^{-4}	21.39	35
5 [b]	FO1	1.73×10^{-2}	19.93	49
6 [b]	FO2	9.97×10^{-5}	19.31	27

[a] Values are taken from reference [15]. [b] Values are taken from reference [14].

From Figure 1 and Table 1, it can be seen that FBS showed better surface activities than FIS, caused by the chemical bond structure, which causes the compounds to have a more stable arrangement at the air–water interface, and thus the A_{CMC} of FBS is 1.65 Å^2/mol, while the A_{CMC} of FIS is 2.46 Å^2/mol.

3.2. Foaming Properties

Comparisons of surface properties with the same hydrophobic fluorinated tail structure compounds (see Figure 2) are summarized in Table 1. The A_{CMC} values of FCS/FAS/FO1/FO2 were larger than those of FIS and FBS, meaning that the technique of introducing an acrylate group into the hydrophilic head group can effectively improve the compact arrangement of molecules at the gas–liquid interface. This causes these fluorosurfactants to have improved application performance in fire-fighting foams.

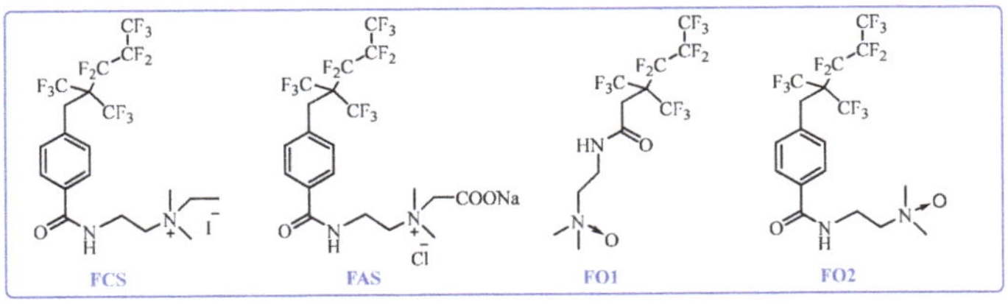

Figure 2. Structures of FCS/FAS/FO1/FO2.

The foaming ratio and eluting time are important indexes used to evaluate the foaming properties of surfactants, representing the foaming ability and foam stability of surfactants, respectively. In this work, we chose a simple method to evaluate the foam properties of the new hydrophilic structure [21]. FCS and FIS were both prepared as 1 wt% aqueous solutions, and 10 mL of these aqueous solutions was added to a 100 mL measuring cylinder with a lid. Then, the cylinder was shaken violently for 20 s to cause the solution to foam fully. From Figure 3A, it can be seen that the initial foam height (4 s after shaking) of FIS was about 85 mL, while the initial foam height of FCS was about 80 mL. This means that the foaming ratios of FIS and FCS are 8.5 times and 8 times, respectively. Meanwhile, the eluting time can be assessed using the time taken for the liquid to precipitate. Figure 3B shows that, at the bottom of both cylinders, about 8 mL of the solution is present at 135 s (after shaking for 115 s), which means that the foam stability of FIS and FCS is almost the same. The above test results show that the more compact arrangement of FIS molecules at the gas–liquid interface improves the foaming ability and has no impact on the foam's stability.

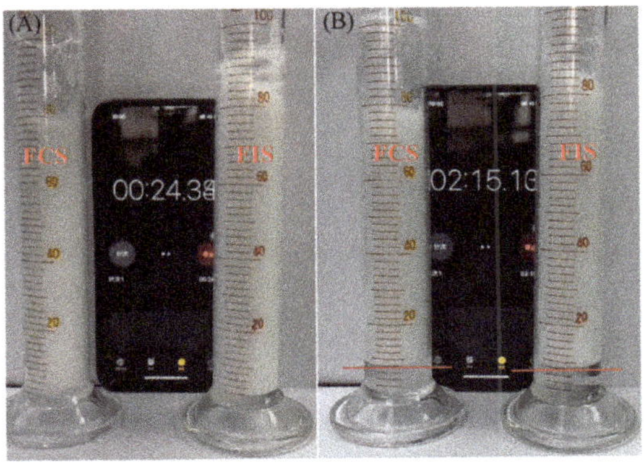

Figure 3. **(A)** Foam height after shaking for 4 s. **(B)** Foam height after shaking for 115 s.

3.3. Coating Applications

The introduction of an acrylate group provides these novel fluorosurfactants with more new application possibilities compared with the traditional fluorinate surfactant. For example, polymerizable emulsifiers are a kind of new emulsifier that do not migrate to the film surface during the film formation process, thus avoiding deleterious effects on the film's properties [22]. We used FBS as a polymerizable emulsifier to discuss its application in the coating industry.

For emulsion 1, a polymerizable emulsifier named DNS-86 purchased from Guangzhou Shuangjian Trading Co., Ltd, Guangzhou, China, was used as a single emulsifier. Meanwhile, for emulsion 2, FBS was used with DNS-86 with a mass radio of 1:1 as a mixed emulsifier. The other additives and total feed weight of emulsion 1 and emulsion 2 were kept consistent. The preparation details of the two kinds of polyacrylate emulsions and the coating process are presented in the Supporting Information, while the general processes are shown in Figure 4A. Transmission electron microscope (TEM) images of the latexes were obtained by a JEOL JEM 1230 instrument operated at 80 kV. The TEM micrograph reveals that both prepared emulsions have a good dispersity, with nanoscale sizes of less than 100 nm (Figure 4B), indicating that small amounts of FBS did not significantly change the micromorphology of the emulsion.

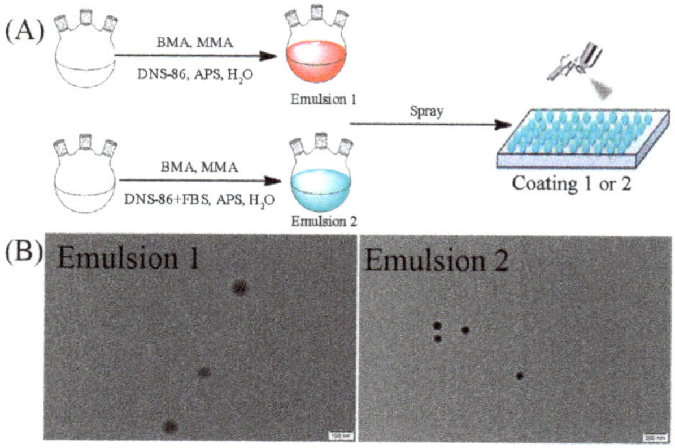

Figure 4. (**A**) Fabrication process of polyacrylate emulsion coatings. (**B**) TEM image of emulsion 1 and emulsion 2.

The contact angle (CA) test is commonly used as a quick and precise tool for the evaluation of the liquid repellence of a coating surface. The measurements were performed on a contact angle goniometer (XG-CAM, Shanghai Xuanyichuangxi Industrial Equioment Co., Ltd., Shanghai, China) by the sessile drop method with a microsyringe at room temperature. We chose a glass slide as a template substrate to perform the water contact angle (WCA) test and hexamethylene contact angle (HCA) test (Figure 5A,B, respectively), and it can be observed that the WCA of the film decreased by only 1% (from 86.63° to 85.42°) with the addition of a small amount of FBS (1.22 wt% in total). Meanwhile, the HCA of the coating improved by over 100% (from 5.02° to 10.13°), caused by the fluorinated groups and quaternary ammonium salt groups of the 1.22 wt% FBS, improving the oleophobic properties of the coating together. Regarding the water repellence of the coatings, the hydrophilic quaternary ammonium salt groups play a dominant role compared with the hydrophobic fluorinated groups.

To assess the remarkable influence of the quaternary ammonium salt groups on the coatings, we further conducted antibacterial tests of the two emulsions by the inhibition zone method according to the literature [23,24]. Figure 6 shows the inhibition zone photos of different bacteria in Petri dishes after bacterial cultures for 96 h; the larger inhibition zone means a better ability to inhibit the corresponding bacteria. From Figure 6, it can be found that the antibacterial zone of emulsion 2 was larger than that of emulsion 1, meaning that the addition of FBS can greatly improve the anti-Staphylococcus aureus activity by approximately 2 times after 96 h and improves the anti-Escherichia coli activity by approximately 1.6 times after 96 h. Unfortunately, both of the latex emulsions showed no anti-Aspergillus niger activity. This phenomenon is mainly due to the presence of the quaternary ammonium salt structure, but the relative amount of FBS in emulsion 2 was too low (below 3 wt‰) that it has not yet acted as an inhibitor against Aspergillus niger.

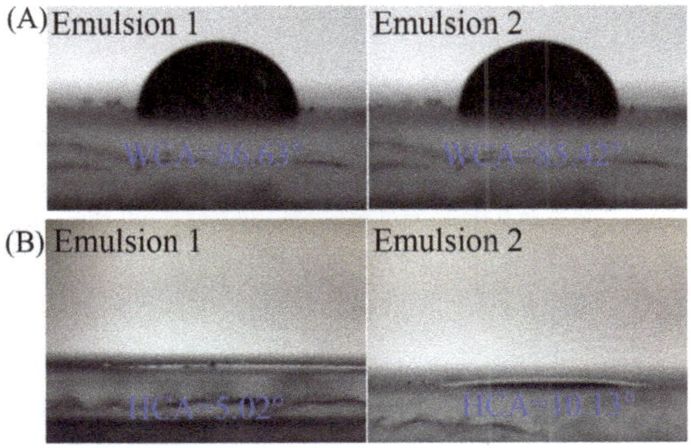

Figure 5. (**A**) Water contact angle (WCA) images of fabricated glasses of emulsion 1 and emulsion 2. (**B**) Hexamethylene contact angle (HCA) images of fabricated glasses of emulsion 1 and emulsion 2.

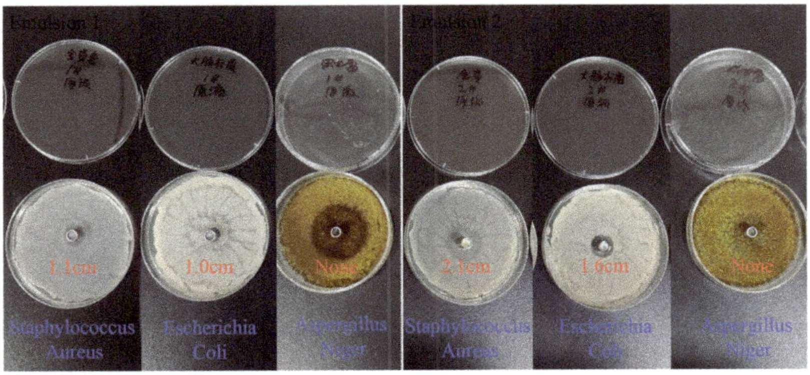

Figure 6. Antibacterial test images of emulsion 1 and emulsion 2 after 96 h.

4. Conclusions

In this work, we successfully synthesized two novel branched fluorosurfactants with multi-functional groups starting from perfluoro-2-methyl-2-pentene. The introduction of an acrylate group into the hydrophilic head group effectively promoted the compact arrangement of molecules at the air–water interface, thus improving the foaming ability of the molecular solutions. The as-obtained compounds not only exhibited better surface properties than those of sodium perfluorooctanoate but also showed remarkable application possibilities compared to the traditional fluorinate surfactant. When FBS was applied as a polymerizable emulsifier, the as-obtained latex emulsion showed much better organic solvent repellence, *anti-Staphylococcus aureus* activity, and *anti-Escherichia coli* activity.

Supplementary Materials: The following supporting information can be downloaded at: https://www.mdpi.com/article/10.3390/surfaces7040055/s1, File S1: synthesis of FIS and FBS; preparation of the polyacrylate emulsion; ^{1}H NMR, ^{19}F NMR and ^{13}C NMR spectra of FIS and FBS.

Author Contributions: Conceptualization, C.L. and B.J.; methodology, P.X.; validation, C.L., J.L. and Y.Q.; formal analysis, C.L.; investigation, C.L.; resources, P.X.; data curation, C.L. and J.L.; writing—original draft preparation, C.L.; writing—review and editing, P.X.; visualization, C.L.; supervision, P.X.; project administration, B.J.; funding acquisition, B.J. All authors have read and agreed to the published version of the manuscript.

Funding: This research was funded by the Science and Technology Commission of Shanghai Municipality, grant number 20DZ1201000.

Institutional Review Board Statement: Informed consent was obtained from all subjects involved in this study.

Data Availability Statement: The original contributions presented in the study are included in the article/Supplementary Material, further inquiries can be directed to the corresponding authors.

Conflicts of Interest: Chao Lin is employed by the Shanghai Fire Research Institute of MEM; Jinhua Li, Yejun Qin, Ping Xing, and Biao Jiang are employed by the Shanghai Institute of Organic Chemistry, Chinese Academy of Sciences. The remaining author declares that this study was conducted in the absence of any commercial or financial relationships that could be construed as potential conflicts of interest.

References

1. Simons, J.H. Electrochemical Process of Making Fluorine-Containing Carbon Compounds. US2519983A, 22 August 1950.
2. Chen, Y.; Petkov, J.T.; Ma, K.; Li, P.; Webster, J.R.; Penfold, J.; Thomas, R.K.; Allgaier, J.; Dalgliesh, R.; Smith, G. Manipulating the hydrophilic/hydrophobic balance in novel cationic surfactants by ethoxylation: The impact on adsorption and self-assembly. *J. Colloid Interface Sci.* **2024**, *674*, 405–415. [CrossRef] [PubMed]
3. da Silva, M.d.G.C.; da Silva, M.E.P.; de Medeiros, A.O.; Meira, H.M.; Sarubbo, L.A. Surface Properties and Biological Activities on Bacteria Cells by Biobased Surfactants for Antifouling Applications. *Surfaces* **2022**, *5*, 383–394. [CrossRef]
4. Drakesmith, F.; Hughes, D. The electrochemical fluorination of octanoyl chloride. *J. Appl. Electrochem.* **1979**, *9*, 685–697. [CrossRef]
5. Haszeldine, R. Synthesis of fluorocarbons, perfluoroalkyl iodides, bromides and chlorides, and perfluoroalkyl grignard reagents. *Nature* **1951**, *167*, 139–140. [CrossRef] [PubMed]
6. Hutchinson, J. The chemistry and properties of a new range of fluorochemical surface active agents. *Eur. J. Lipid Sci. Technol.* **1974**, *76*, 158–163. [CrossRef]
7. Murphy, P.M.; Baldwin, C.S.; Buck, R.C. Syntheses utilizing n-perfluoroalkyl iodides [R_FI, C_nF_{2n+1}-I] 2000–2010. *J. Fluor. Chem.* **2012**, *138*, 3–23. [CrossRef]
8. Ng, C.A.; Hungerbuehler, K. Exploring the Use of Molecular Docking to Identify Bioaccumulative Perfluorinated Alkyl Acids (PFAAs). *Environ. Sci. Technol.* **2015**, *49*, 12306–12314. [CrossRef]
9. Renner, R. The long and the short of perfluorinated replacements. *Environ. Sci. Technol.* **2006**, *40*, 12–13. [CrossRef]
10. Venkatesan, A.K.; Halden, R.U. Loss and in situ production of perfluoroalkyl chemicals in outdoor biosolids–soil mesocosms. *Environ. Res.* **2014**, *132*, 321–327. [CrossRef]
11. Vierke, L.; Möller, A.; Klitzke, S. Transport of perfluoroalkyl acids in a water-saturated sediment column investigated under near-natural conditions. *Environ. Pollut.* **2014**, *186*, 7–13. [CrossRef]
12. Lu, Y.; Gao, K.; Li, X.; Tang, Z.; Xiang, L.; Zhao, H.; Fu, J.; Wang, L.; Zhu, N.; Cai, Z.; et al. Mass Spectrometry-Based Metabolomics Reveals Occupational Exposure to Per- and Polyfluoroalkyl Substances Relates to Oxidative Stress, Fatty Acid β-Oxidation Disorder, and Kidney Injury in a Manufactory in China. *Environ. Sci. Technol.* **2019**, *53*, 9800–9809. [CrossRef] [PubMed]
13. Zhong, H.; Liu, W.; Li, N.; Ma, D.; Zhao, C.; Li, J.; Wang, Y.; Jiang, G. Assessment of perfluorohexane sulfonic acid (PFHxS)-related compounds degradation potential: Computational and experimental approaches. *J. Hazard. Mater.* **2022**, *436*, 129240. [CrossRef]
14. Lin, C.; Pan, R.; Xing, P.; Jiang, B. Synthesis and combined properties of novel fluorinated cationic surfactants derived from hexafluoropropylene dimer. *Chin. Chem. Lett.* **2018**, *29*, 1613–1616. [CrossRef]
15. Sha, M.; Pan, R.; Xing, P.; Jiang, B. Synthesis and surface activity study of branched fluorinated cationic (FCS), gemini (FGS) and amphoteric (FAS) surfactants with $CF_3CF_2CF_2C(CF_3)_2$ group. *J. Fluor. Chem.* **2015**, *169*, 61–65. [CrossRef]
16. Zhang, D.; Sha, M.; Pan, R.; Lin, X.; Xing, P.; Jiang, B. Design and synthesis of the novel branched fluorinated surfactant intermediates with $CF_3CF_2CF_2C(CF_3)_2$ group. *Chin. Chem. Lett.* **2019**, *30*, 566–568. [CrossRef]
17. Kostov, G.; Boschet, F.; Ameduri, B. Original fluorinated surfactants potentially non-bioaccumulable. *J. Fluor. Chem.* **2009**, *130*, 1192–1199. [CrossRef]
18. Zhang, Q.; Luo, Z.; Curran, D.P. Separation of "light fluorous" reagents and catalysts by fluorous solid-phase extraction: Synthesis and study of a family of triarylphosphines bearing linear and branched fluorous tags. *J. Org. Chem.* **2000**, *65*, 8866–8873. [CrossRef]
19. Holt, D.J.; Payne, R.J.; Chow, W.Y.; Abell, C. Fluorosurfactants for microdroplets: Interfacial tension analysis. *J. Colloid Interface Sci.* **2010**, *350*, 205–211. [CrossRef] [PubMed]
20. Ngo, T.H.V.; Damas, C.; Naejus, R.; Coudert, R. A comparative study of micellization with fluorinated and hydrogenated diquaternary ammonium bolaamphiphiles. *J. Colloid Interface Sci.* **2013**, *400*, 59–66. [CrossRef]
21. Xiao, J.; Gao, Z.; Wang, M.; Cui, R. Laboratory method for evaluating the performance of aqueous film-forming foam fire-extinguishing agent. *Chem. Res. Appl.* **2008**, *20*, 569–572.
22. Zhou, J.; Chen, X.; Duan, H.; Ma, J. Synthesis and characterization of organic fluorine and nano-SiO_2 modified polyacrylate emulsifier-free latex. *Prog. Org. Coat.* **2015**, *89*, 192–198. [CrossRef]

3. Li, B.; Li, Y.; Zhao, Y.; Sun, L. Shape-controlled synthesis of Cu_2O nano/microcrystals and their antibacterial activity. *J. Phys. Chem. Solids* **2013**, *74*, 1842–1847. [CrossRef]
4. Cooper, K. The theory of antibiotic inhibition zones. In *Analytical microbiology*; Elsevier: Amsterdam, The Netherlands, 1963; pp. 1–86.

surfaces

MDPI

Article

Charged Microdroplets Deposition for Nanostructured-Based Electrode Surface Modification

Rosaceleste Zumpano [1,*], Marco Agostini [1], Franco Mazzei [1], Anna Troiani [1], Chiara Salvitti [1], Marta Managò [1], Alessia Di Noi [1], Andreina Ricci [2] and Federico Pepi [1,*]

[1] Department of Chemistry and Drug Technologies, "Sapienza" University of Rome, P.le Aldo Moro 5, 00185 Rome, Italy; marco.agostini@uniroma1.it (M.A.); franco.mazzei@uniroma1.it (F.M.); anna.troiani@uniroma1.it (A.T.); chiara.salvitti@uniroma1.it (C.S.); marta.manago@uniroma1.it (M.M.); alessia.dinoi@uniroma1.it (A.D.N.)

[2] Department of Mathematics and Physics, University of Campania L. Vanvitelli, Viale Lincoln 5, 81100 Caserta, Italy; andreina.ricci@unicampania.it

* Correspondence: rosaceleste.zumpano@uniroma1.it (R.Z.); federico.pepi@uniroma1.it (F.P.)

Abstract: Accelerated synthesis of gold nanoparticles (AuNPs) in charged microdroplets produced by electrospray ionization (ESI) was exploited to modify the surface of graphite screen-printed electrodes (GSPEs). The deposited AuNPs were then functionalized by the charged microdroplets deposition of 6-ferrocenyl-hexanethiol (6Fc-ht) solutions that act as reducing and stabilizing agents and provide electrochemical properties for the modified electrodes. The morphology and composition of the AuNPs were characterized by scanning electron microscopy (SEM). Cyclic voltammetry (CV), differential pulse voltammetry (DPV) and electrochemical impedance spectroscopy (EIS) were used to investigate the electrochemical behavior of the modified electrodes. The results showed that the ESI microdroplets deposition technique produces uniform and well-dispersed AuNPs on GSPE, and optimal conditions for deposition were identified, enhancing GSPE electrocatalytic performance. Further functionalization by ESI microdroplets of AuNPs with 6Fc-ht demonstrated improved redox properties compared with the conventional self-assembled monolayer (SAM) method, highlighting the technique's potential for the easy and fast functionalization of electrochemical sensors.

Keywords: microdroplets; ESI Z-spray; screen-printed electrodes; gold nanoparticles

1. Introduction

Electrospray ionization (ESI) is a mass spectrometric technique that allows obtaining gaseous naked ions of nonvolatile molecules by spraying microdroplets into a high-voltage electric field whose invention by John Fenn was awarded the Nobel Prize [1]. In recent years, ESI microdroplets have gained additional interest in the scientific community as they have been demonstrated to be a unique medium where ionic reactions can be accelerated 10^6 times with respect to the same bulk processes [2]. Starting from the first studies by Cooks and Zare [3,4], a large variety of chemical and biological processes in solution were demonstrated to occur in the microseconds fly-time of the microdroplets before their Coulombic explosion into naked ions [5–15].

Furthermore, the microdroplets deposited onto a solid surface generate a thin film that retains their peculiar, confined volume but allows for the neutral products of the reaction to be separated and quantified [16–25]. Moreover, the thin film is continuously formed by microdroplets deposition, thus extending the reaction time to any desired longer value with respect to the microsecond droplet lifetime. Furthermore, several studies demonstrated that metal nanoparticles can be efficiently deposited onto a target surface from solutions of metal ion salts [26–28]. These unusual applications of mass spectrometry have enormously broadened the field of its applications from a purely analytical technique to a procedure capable of supporting or improving classic synthetic methods in solution [29,30]. Gold nanoparticles (AuNPs) are nanoscale gold particles with unique optical, electrical, and catalytic properties [31]. One of the challenges in using AuNPs for various applications is to control their size, shape, and surface chemistry. Functionalization of AuNPs with organic molecules such as thiols can modify their surface properties and enhance their stability, biocompatibility, and functionality. Recently, Zare and coworkers demonstrated that AuNPs can be formed by ESI microdroplets deposition at ambient conditions by spraying a solution containing only tetracholoroauric acid without adding reducing agents [32].

Moreover, AuNPs of less than 10 nm in diameter were generated in the microsecond time scale. The fast formation of AuNPs was attributed to the strong electric field at the water–air microdroplets interface. In this study, the procedure developed by Zare was used to modify the surface of graphite screen-printed electrodes (GSPE). The morphology of the AuNPs was characterized by scanning electron microscopy (SEM), whereas cyclic voltammetry (CV), differential pulse voltammetry (DPV) and electrochemical impedance spectroscopy (EIS) were used to investigate the electrochemical behavior of the modified electrodes. Furthermore, the same deposition method was used to functionalize the AuNPs-modified electrode surface with 6-ferrocenyl-hexanethiol (6Fc-ht) as a redox probe. The results obtained pave the way for using this deposition technique to develop electrochemical sensors and biosensors.

2. Materials and Methods

2.1. Reagentes and Samples

Potassium ferricyanide ($K_3[Fe(CN)_6]$), potassium ferrocyanide ($K_4[Fe(CN)_6]$), potassium chloride (KCl), tetrachloroauric acid ($HAuCl_4$), 6-ferrocenyl-hexanethiol (6Fc-ht), and methanol (MeOH) were purchased by Merck Life Science (Milan, Italy). All solutions used for the electrochemical measurements were prepared using Milli-Q water (R = 18.2 MΩ cm at 25 °C; TOC < 10 μg L^{-1}, Millipore, Molsheim, France). GSPE electrodes were purchased by Metrohm Italiana S.r.L. (Origgio, Italy) $K_3[Fe(CN)_6]/K_4[Fe(CN)_6]$ 1 mM and 5 mM solutions were prepared in KCl 0.1 M. The 6-Ferrocenyl-hexanethiol 1 mM stock solution used to modify GSPE electrodes via ESI and SAM methods was prepared by dissolving the reagent in pure MeOH.

2.2. ESI Z-Spray Microdroplets Deposition Experiments

Microdroplets deposition experiments were performed by using the Z-spray ionization (ESI) source of a quadrupole-time of flight (Q-TOF, Ultima, Micromass, Manchester, UK) mass spectrometer suitably adapted to microdroplets reaction studies [33]. Briefly, in the ESI Z-spray source, the microdroplets, dried by the N_2 desolvation gas, hit the GSPE working surface fixed on a target plate whose distance from the exit of the ESI capillary can be varied from 1.0 to 3.0 cm. The GSPE working surface was biased with the same voltage applied to the ESI source cone, thus allowing the charged microdroplets stream to be selectively focalized onto this part of the screen-printed electrode (supplementary materials, Figure S1). The sprayed $HAuCl_4$ water solutions were prepared daily at a total concentration of 6.4×10^{-5} M. Nitrogen was used as desolvation gas at a flow rate of 200 L h^{-1} and at a temperature of 200 °C (corresponding to an actual microdroplets stream temperature of 70 °C). The ESI Z-spray experiments involving $HAuCl_4$ were performed in negative ion mode. Typical source potentials were as follows: capillary −2.5 kV, cone voltage 70 V, RF lens-1 120 V, and syringe pump flow 20 μL min^{-1}. After the AuNPs deposition, the GSPE surfaces were treated with a spray of microdroplets containing 6Fc-ht. The 6Fc-ht methanol solutions were prepared daily at a total concentration of 1.0×10^{-5} M and were delivered to the GSPE working surface for 10 min. The 6Fc-ht deposition through the ESI Z-spray source was performed in positive ion mode. Typical source potentials were as follows: capillary 3.5 kV, cone 70 V, RF lens-1 120 V, desolvation gas flow rate 200 L h^{-1}, temperature 200 °C and syringe pump flow 20 μL min^{-1}. In the case of H_2O_2 oxidation catalysis, the DPV measurements were conducted in a 5 mL solution of PBS 0.01 M and

KCl 0.1 M by adding increasing amounts of H_2O_2 from a 5 mM mother solution prepared in PBS 0.01 M.

2.3. Surface Characterization Apparatus

Scanning electron microscopy (SEM) was performed to characterize the electrode surface morphology using a Dual Beam Auriga (Zeiss instrument, Oberkochen (Germany) of the Sapienza Nanoscience and Nanotechnology Labs (SNN-Lab). Elemental evaluation of the AuNPs was ascertained by energy-dispersive X-ray (EDX) analysis.

2.4. Electrochemical Measurements and Apparatus

Cyclic voltammetry (CV) and differential pulse voltammetry (DPV) measurements were conducted in a three-electrode system consisting of a platinum wire as the counter electrode and an $Ag/AgCl_{sat.}$ electrode as reference. The experiments were performed in a 10 mL solution of 1 mM $K_3[Fe(CN)_6]/K_4[Fe(CN)_6]$ with KCl 0.1 M as a supporting electrolyte. Electrochemical impedance spectroscopy (EIS) measurements were carried out using the same three-electrode setup in a 5 mM $K_3[Fe(CN)_6]/K_4[Fe(CN)_6]$ solution with KCl 0.1 M as supporting electrolyte. The experiments involving 6Fc-ht, CV, and DPV measurements were performed in 0.1 M KCl. All experiments were conducted at room temperature.

2.5. Self-Assembled Monolayer (SAM) Formation

The procedure to provide the formation of a SAM of 6Fc-ht onto the GSPE/AuNPs electrodes was realized by dipping the modified electrodes in a 1 mM thiol solution in ethanol for 24 h. Afterward, the electrodes were left to dry and rinsed with water and ethanol.

3. Results and Discussion

3.1. ESI Microdroplets Deposition Experiments

To probe the formation of gold in the Q-TOF ESI Z-spray modified source, a stream of aqueous microdroplets from a 6.4×10^{-4} M $HAuCl_4$ solution was collected in negative ion mode onto a glass slide for 1 h. The formation of a naked-eye visible gold film clearly evidenced the capability of our systems to reproduce the Zare experiments (Figure S1). Moreover, even the negative ion mass spectrum registered with the Q-TOF mass-spectrometer was comparable with that reported by Zare et al., showing the signal corresponding to $AuCl_2^-$ and $AuCl_4^-$ as the prevalent anionic species produced by the complete desolvation of the ESI microdroplets. As underlined by Zare, the $AuCl_2^-$ species clearly indicates that Au^{+3} is reduced inside the microdroplets. With the aim to promote the deposition of gold nanoparticles with as small as a nanometric diameter, the tetrachloroauric acid solution was diluted ten times (6.4×10^{-5} M), and the reaction time was reduced in the range of 2–10 min. The selected deposition times are considerably longer than the millisecond interval used by Zare et al. in their experiments. This choice was necessary to obtain intense electrochemical signals, but it led to an increase in the size of gold nanoparticles, which underwent aggregation phenomena (vide infra). The ESI microdroplets were collected onto the working area of a GSPE. ESI source parameters, such as electric field polarity and spray-target distance, may strongly affect the microdroplets deposition processes [34,35].

In a series of preliminary experiments, the distance between the ESI spray tip and the specific portion of the GSPE that has to be modified by AuNPs was varied from 1.0 to 3.5 cm. The DPV analysis of the AuNPs modified electrodes suggests that the optimal electrochemical signal was associated with the deposition of the ESI microdroplet stream at a distance to the graphite electrode of 2.5 cm. This parameter was then used to modify the working area of the GSPE electrodes with AuNPs in all the subsequent experiments. Three experiments were conducted to determine the efficiency of the gold nanoparticle deposition process by extending the deposition time to 150 min to reach weighable quantities of gold onto the GSPE surface. The modified electrodes were washed with water to remove chloroauric acid residues and placed in an oven at 100 °C for 1 h. The measured electrode weight increase corresponds to an average conversion of 60% of the total chloroauric acid delivered to the surface. The AuNPs-modified electrodes were then functionalized by the deposition of microdroplets containing 6Fc-ht as a redox mediator. The only difference in the mass spectrometric condition used in these experiments was the ESI voltage polarity that was switched to positive values, leading to better results in terms of DPV and CV measurements. The positive ESI mass spectrum of 6Fc-ht is dominated by a single signal corresponding to its molecular ion at m/z 302 (Figure S2).

3.2. SEM and Electrochemical Characterization of the AuNPs-Modified Electrodes

The morphology and distribution of AuNPs obtained through the ESI microdroplets deposition after 2, 4 and 10 min of an $HAuCl_4$ solution have been analyzed by SEM microscopy (Figure 1). In the Supplementary Materials (Figure S3), the EDX analysis is reported, confirming the presence of gold on the electrode surface.

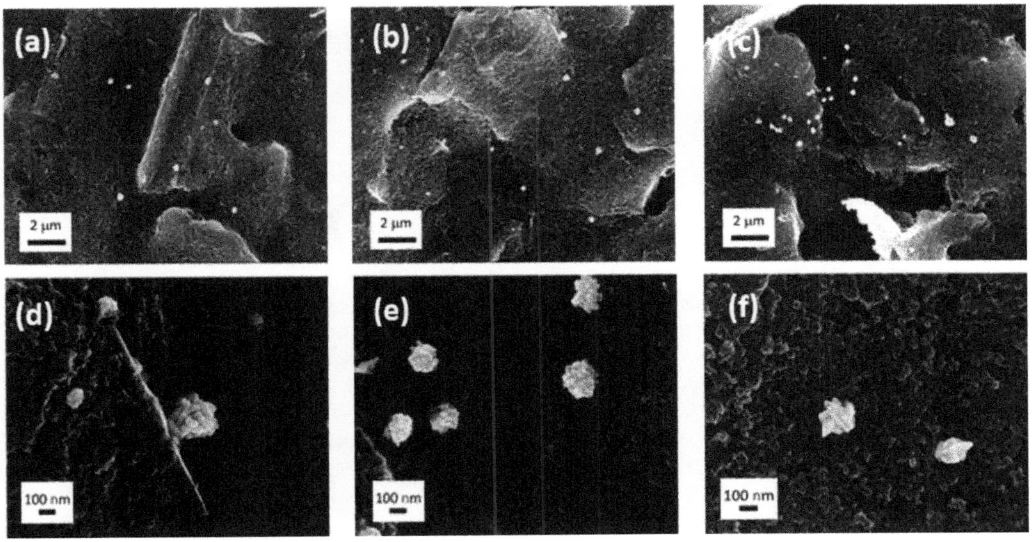

Figure 1. SEM images: (**a**,**d**) 2 min of deposition; (**b**,**e**) 4 min of deposition; (**c**,**f**) 10 min of deposition.

The total amount of $HAuCl_4$ delivered to the electrode surface increases by increasing the deposition time as well (2 min: 0.00087 mg; 4 min: 0.0017 mg; 10 min: 0.0043 mg). Between 2 and 4 min, the AuNPs number accordingly increases, and the distribution over the surface appears homogeneous after 4 min. After 10 min of deposition, the AuNPs start to concentrate more in specific regions (Figure 1a–c). The AuNPs morphology results in large anisotropic aggregates of smaller NPs (of approximately 30 nm) that are not completely formed at 2 min of deposition (Figure 1d and Figure S4) with the co-presence of AuNPs of approximately 350 nm and 100 nm.

After 4 min (Figure 1e), the medium size for all the AuNPs aggregates is approximately 220 nm, and the aggregates are formed by smoother NPs (Figure S5). After 10 min of deposition, the medium size is approximately 260 nm with more spikey morphology (Figure 1c,f and Figure S6). In addition, the presence of random bigger aggregates suggests the occurrence of successive nucleations during the longer deposition time.

The graphics of the statistical distribution of AuNPs content and size for three different batches at 2, 4, and 10 min of deposition are reported in Figures S8–S10.

The electrochemical characterization of GSPE/AuNPs platforms was realized using CV, DPV, and EIS measurements. The CV curves (Figure 2a) show increasing oxidation and reduction peak currents for the redox couple $K_4[Fe(CN)_6]/K_3[Fe(CN)_6]$ with increasing $HAuCl_4$ deposition time, as well as the reversibility of the reaction. These results are consistent with the widely known property of AuNPs to promote ET efficiency, increasing the surface-to-volume ratio, as well as the electrode conductivity [31,36].

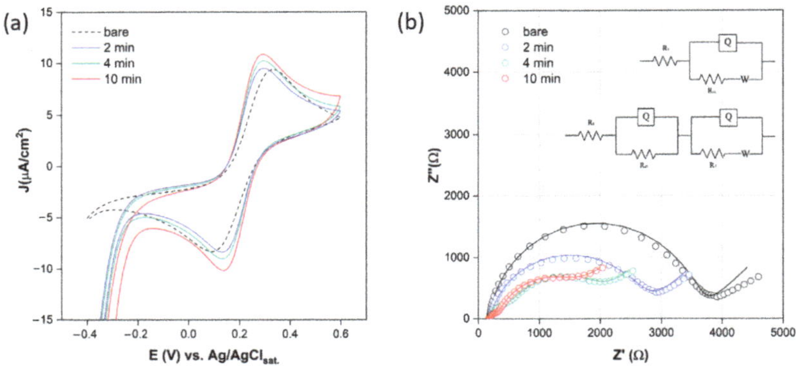

Figure 2. (**a**) CV measurements of AuNPs-GSPE surfaces modified under different deposition times, performed in $K_4[Fe(CN)_6]/K_3[Fe(CN)_6]$ 1 mM, KCl 0.1 M, between −0.4 and 0.6 V vs. $Ag/AgCl_{sat.}$, rate = 10 mV/s. (**b**) EIS measurements of AuNPs-GSPE surfaces under different deposition times, performed in $K_4[Fe(CN)_6]/K_3[Fe(CN)_6]$ 5 mM, KCl 0.1 M, E_{DC} = 0.215 V, E_{AC} = 0.01 V, νmax = 100,000 Hz, νmin = 0.1 Hz. The dotted lines represent experimental data, and the solid lines represent the fit lines. Inset: Randles (R(Q[RW])) and R(QR)(Q[RW]) circuits employed for the fitting procedure.

The EIS measurements are carried out for the same platforms, and the Nyquist plots are reported in Figure 2b. The impedance curves are fitted using a typical Randles circuit in the case of 0 and 2 min, while those at 10 min are fitted using an R(QR)(Q[RW]) circuit, obtaining the characteristic system parameters reported in the Supplementary Materials (Table S1).

All the platforms are semi-reversible systems, according to the relationship $R_{CT} \geq R_W = \sigma/\Omega$, which is the condition for having a visible kinetic region over the mass transfer region [37]. By increasing the deposition time, the R_{CT} decreases and, interestingly, at 4 and 10 min, an additional small semi-circle at higher frequencies (ν) appears. This resistance and capacitance are attributable to the AuNPs interface (AuI), which becomes visible at higher AuNPs concentrations [38].

However, the presence of AuNPs, besides increasing the electrode conductivity, confers an intrinsic heterogeneity to the electrode surface, which is responsible for a non-uniform formation of the double layer. Therefore, the capacitance contribution decreases, as well as both the τ_{CT} and τ_d. The Bode plots reported in Figure 3a,b show the variation of $|Z|$ and phase angle (ϕ), respectively, with increasing applied ν. The plateau region under high ν in the $|Z|$ profile results in ν-independencet for all the modified electrodes, following the typical resistive behavior under such conditions [39].

The slope relies on the system's capacitance, which is responsible for the progressively higher phase shift between the current signal and the applied voltage. The greater the phase shift, the poorer the current flow through the system. After 2 min of deposition, $|Z|$ decreases relative to the bare electrode across the entire applied ν range, and ϕmax decreases, shifting slightly to lower ν. This indicates that the system still exhibits capacitive dominance. With increased deposition times of 4 and 10 min, both $|Z|$ and ϕ profiles decrease in magnitude compared to the bare and 2 min electrodes and show a splitting into two distinct phenomena, reflecting the Nyquist plot behavior. The phenomenon at intermediate ν represents the AuI contribution, while the one at lower ν is related to the double layer. After 10 min, despite the decrease in $|Z|$ magnitude, the ϕ_{max} related to the double-layer capacitance increases and shifts to lower ν. This is likely due to the formation of a more uniform double layer as a result of reaching a concentration of AuNPs that promotes increased double-layer capacitance compared to the 4 min electrode. Additionally, the increasing slope of the |Z| profile at low ν is notable, which can be attributed to the progressive decrease in τ_{CT}, making diffusion the rate-limiting step of the process [40].

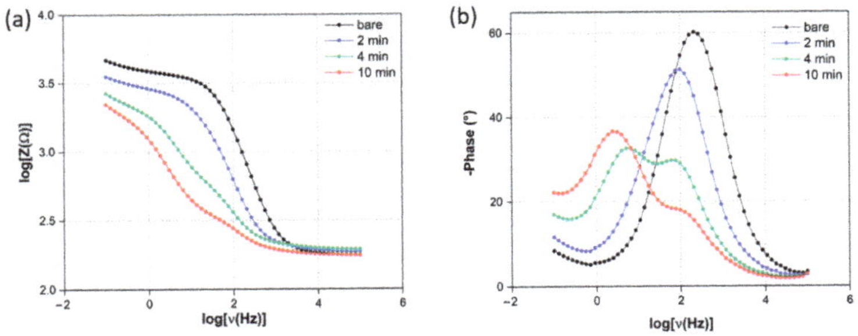

Figure 3. Bode plots (**a**) |Z| and (**b**) phase angle behavior in the function of applied ν.

DPV measurements conducted under oxidation and reduction conditions (Figure 4a,b) allowed for a better characterization of the electrochemical behavior of the modified platforms. For ferrocyanide oxidation, the peak occurs at 0.22 V for the bare electrode, progressively shifting to lower potentials with increasing deposition time (0.2 V at 2 min, 0.15 V at 4 and 10 min). This indicates a gradually more favorable ET process promoted by the increasing concentration of AuNPs. The peak current also increases, reaching its highest value at 10 min. Similar behavior is observed under reduction conditions. After 4 and 10 min, an additional ET process is evident under both oxidation and reduction conditions, even in the absence of oxygen and ferricyanide likely due to the oxidation and reduction of Au. The literature shows that oxidation under acidic conditions is generally not appreciable for adsorbed spherical AuNPs [41], whereas reduction is observed in DPV at 0.45 V vs. $Ag/AgCl_{sat.}$ In this study, the highly anisotropic morphology of the AuNPs could account for their particularly high reactivity. The oxidation peak appears at 0.6 V at 4 min and shifts to 0.49 V at 10 min, with an increase in current intensity. This is probably due to the more spiky morphology [42,43] of AuNPs obtained after 10 min of deposition (Figure 1e,f) and their higher concentration. In the case of reduction, the additional ET process is less pronounced, increasing in current with the higher AuNPs concentration.

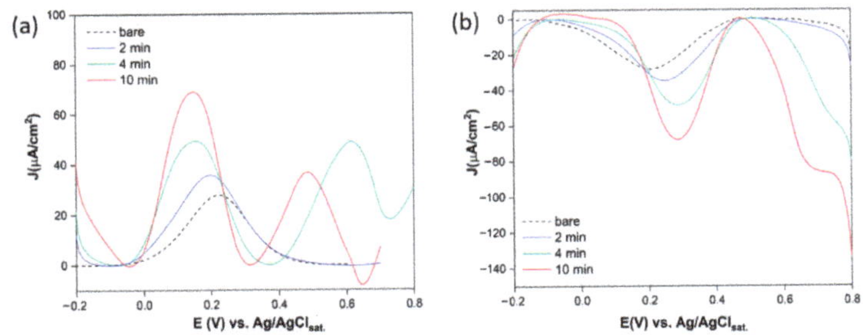

Figure 4. DPV measurements of AuNPs-GSPE surfaces under different deposition times, performed in $K_4[Fe(CN)_6]/K_3[Fe(CN)_6]$ 1 mM, KCl 0.1 M, E_{DC} = 0.215 V, E_{AC} = 0.01 V, νmax = 100,000 Hz, νmin = 0.1 Hz.

3.3. *Functionalization of GSPE/AuNPs Platforms with 6-Ferrocenyl-Hexanethiol*

Finally, the GSPE/AuNPs platforms characterized above have been functionalized with 6Fc-ht as a redox mediator. The functionalization was realized by the conventional self-assembled monolayer (SAM) formation (see Section 2.5) and by the ESI microdroplets deposition technique. In both cases, prior to the characterization, the electrodes were rinsed with abundant water and ethanol to remove the excess of physically adsorbed 6Fc-ht.

The CV measurements have also been carried out for these systems (Figure 5a,b). The system modified by the SAM procedure did not provide appreciable oxidation and reduction signals for all the $HAuCl_4$ deposition times. This is probably due to strongly adsorbed redox probe molecules onto free graphite through the aliphatic chain, inducing high capacitive current and ferrocene's limited mobility. Consequently, only a weak ferrocene oxidation signal can barely be observed in Figure 5a (1, 2, 3) at higher potentials than expected [44]. Gold broad oxidation signals are also detected in Figure 5a (4, 5) at 4 and 10 min.

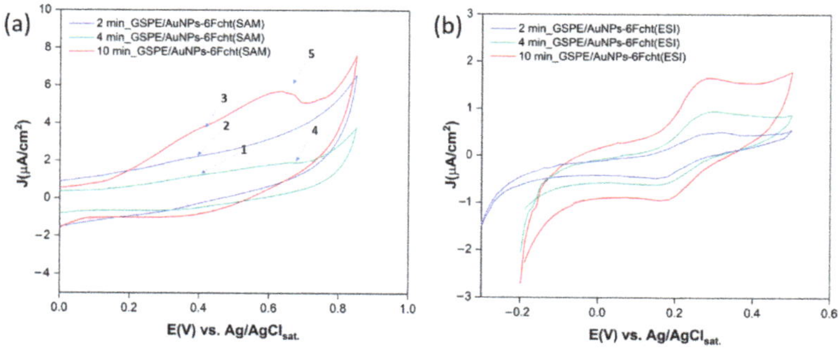

Figure 5. (**a**) CV measurements of GSPE surfaces functionalized with AuNPs at different $HAuCl_4$ deposition times, modified with 6Fc-ht through the SAM procedure performed in KCl 0.1 M, between 0 and 0.85 V vs. $Ag/AgCl_{sat.}$, rate = 10 mV/s. 6Fc-ht oxidation signals (1, 2, and 3) and gold oxidation signals (4, 5); (**b**) CV measurements of GSPE surfaces functionalized with AuNPs at different $HAuCl_4$ deposition times, modified with 6Fc-ht through the ESI technique performed in KCl 0.1 M between −0.2 and 0.5 V vs. $Ag/AgCl_{sat.}$, rate = 10 mV/s.

On the contrary, the ESI deposition technique provided for better CV profiles (Figure 5b) even if the electrode surface was exposed to a less concentrated 6Fc-ht microdroplets stream for only 10 min. The oxidation and reduction peaks are detected at 0.26 and 0.22 V vs. $Ag/AgCl_{sat.}$ respectively, and the current increases with increasing deposition time. These results could be attributed to the peculiar properties of the thin film formed by the charged microdroplets deposition to strongly accelerate the specific interaction between 6Fc-ht and AuNPs with respect to the bulk process. Moreover, the limited quantity of the reagent used to functionalize the gold nanoparticles seems to avoid its physical interaction with the free graphite. In contrast, the SAM procedure exposes the electrode surface to a concentrated solution of 6Fc-ht during a period of 24 h and likely promotes its extensive physical adsorption onto the free graphite.

Finally, the DPV measurements under oxidation and reduction conditions were performed for the systems obtained through different modification procedures (Figure 6).

Although oxidation and reduction currents increase with deposition time for the SAM procedure, the redox process appears irreversible when comparing current intensities. Additionally, the oxidation and reduction potentials significantly deviate from the expected values for ferrocene. These results are consistent with the CV data discussed earlier. In contrast, the DPV for the platforms obtained via the ESI technique shows better reversibility. For all deposition times, oxidation and reduction peaks appear at 0.13 V and 0.12 V vs. $Ag/AgCl_{sat}$, respectively, aligning with the ferrocene redox behavior reported in the literature under similar conditions [45]. Lastly, gold oxidation is consistently observed at 4 and 10 min of deposition, except in the reduction phase of the SAM system.

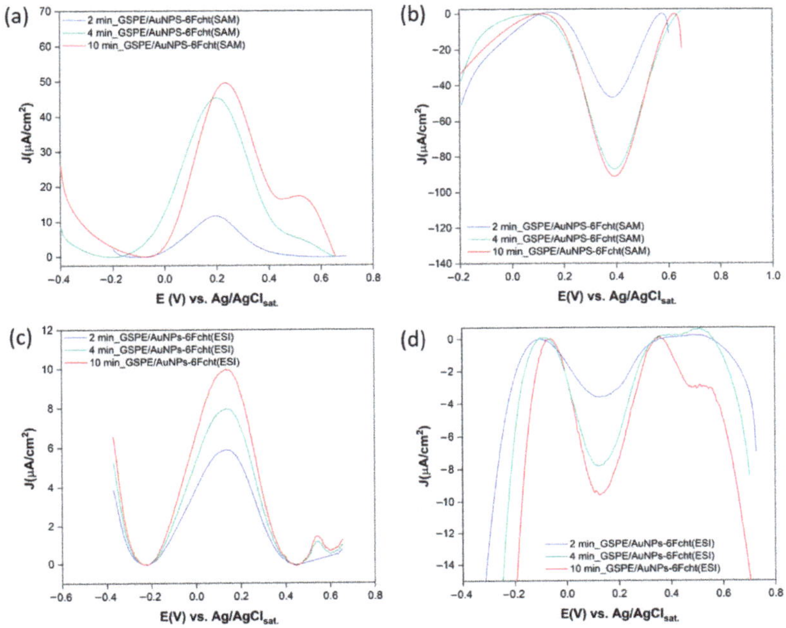

Figure 6. DPV measurements of GSPE surfaces functionalized with AuNPs at different $HAuCl_4$ deposition times, modified with 6Fc-ht through the SAM procedure (**a**,**b**) and ESI microdroplets deposition technique (**c**,**d**), performed in KCl 0.1 M, E_{DC} = 0.215 V, E_{AC} = 0.01 V, νmax = 100,000 Hz, and νmin = 0.1 Hz.

3.4. Electrochemical Detection of H_2O_2

The ESI GSPE/AuNPs-6Fcht platform at 10 min of microdroplets deposition, which provides the best electrochemical performance, was employed for the detection of H_2O_2. Specifically, the catalysis of H_2O_2 oxidation by 6Fc-ht was followed through the DPV measurements (Figure 7), where the Fc oxidation current increases with increasing H_2O_2 concentration, as recently reported in the literature [46].

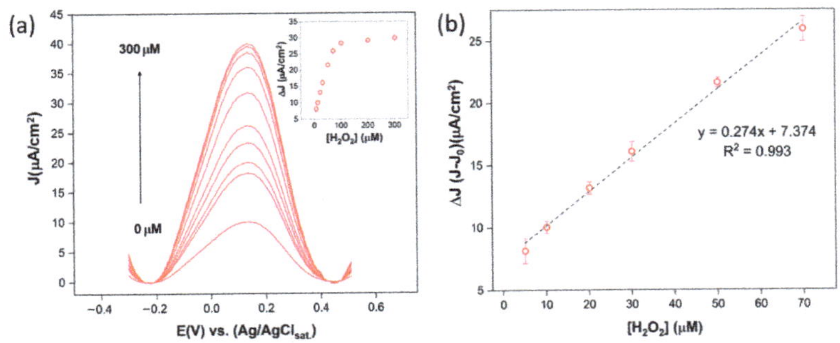

Figure 7. (**a**) DPV curves collected in the absence and presence of increasing concentrations of H_2O_2 in the range 0–300 μM in PBS 0.01 M, KCl 0.1 M. In the inset are reported the ΔI (I-I_0) over the analyzed H_2O_2 concentration range; (**b**) sensor calibration line.

The mechanism proposed in the literature involves the oxidation of Fc to Fc^+ and the catalysis of H_2O_2 oxidation to O_2 by Fc^+. Afterwards, Fc is oxidized back to Fc^+, increasing the peak current proportionally to the H_2O_2 concentration present in the working solution. The platform exhibited an optimal response to the target, providing for a linear range of 5–70 μM with a limit of detection (LOD) of 0.9 μM. These results make this platform promising for applications in reactive oxygen species (ROS) detection.

4. Conclusions

This study demonstrates the potential of the ESI-charged microdroplets deposition technique for forming and functionalizing AuNPs on GSPEs. By optimizing deposition parameters such as ESI voltage polarity and spray-target distance, we achieved a homogenous distribution of AuNPs, verified through SEM analyses. The advantages of the ESI microdroplets deposition technique over the common AuNPs generation methods can be summarized as follows: (i) minor reagents consumption due to the use of $HAuCl_4$ diluted solution in the absence of reducing agents, (ii) accelerated nanoparticle formation rate, (iii) elimination of time-intensive purification steps like centrifugation or filtration. Electrochemical characterization using CV, DPV, and EIS confirmed the enhanced electrocatalytic properties of the AuNPs-modified electrodes, attributed to improved ET efficiency and increased surface conductivity. Functionalization with 6-ferrocenyl-hexanethiol via the same ESI microdroplets deposition technique also resulted in superior redox behavior compared to conventional self-assembled monolayer (SAM) methods. The ESI technique provided specific interaction between 6-ferrocenyl-hexanethiol and AuNPs, promoting effective redox activity while minimizing non-specific adsorption. Moreover, deposition and functionalization of the AuNPs through charged microdroplets deposition may occur in two consecutive steps of a few minutes without any electrode surface manipulation, saving processing time and improving the uniformity and repeatability of the modified layer. Finally, the platform at 10 min of microdroplets deposition showing the best electrochemical performance was used to develop an electrochemical sensor for H_2O_2 detection. The device exhibited a linear range of 5–70 μM and a LOD of 0.9 μM.

These findings highlight the versatility and efficacy of ESI microdroplets deposition in fabricating and functionalizing nanostructured electrodes, paving the way for advanced electrochemical sensors and biosensors.

Supplementary Materials: The following supporting information can be downloaded at: https://www.mdpi.com/article/10.3390/surfaces7040052/s1, Figure S1: ESI Z-spray apparatus, Figure S2 and S3: ESI mass spectra, Figure S4: EDX analysis, Figures S5–S7 Magnified SEM images of AuNPs, Figure S8: Size distribution of AuNPs after 2 min of microdroplets deposition, Figure S9: Size distribution of AuNPs after 4 min of microdroplets deposition, Figure S10: Size distribution of AuNPs after 10 min of microdroplets deposition, Table S1: EIS parameters.

Author Contributions: R.Z. and M.M.: methodology, investigation, formal analysis, writing—original draft. F.M., F.P., R.Z. and M.M.: conceptualization, resources, writing—original draft, review, and editing. M.A., A.T., C.S., A.D.N. and A.R.: resources, review, and editing. All authors have read and agreed to the published version of the manuscript.

Funding: This work was supported by Sapienza Università di Roma "Progetti di Ateneo 2020", RM120172B7339DD4. C.S. thanks the Italian Ministry of University and Research (MUR) for a Researcher position (DM n. 1062, 10/08/2021) within the EU-funded National Operational Programme (PON) on Research and Innovation 2014–2020.

Data Availability Statement: The dataset is available on request from the authors.

Conflicts of Interest: The authors declare that they have no known competing financial interests or personal relationships that could have appeared to influence the work reported in this paper.

References

1. Fenn, J.B.; Mann, M.; Meng, C.K.; Wong, S.F.; Whitehouse, C.M. Electrospray ionization for mass spectrometry of large biomolecules. *Science* **1989**, *246*, 64–71. [CrossRef] [PubMed]
2. Wei, Z.; Li, Y.; Cooks, R.G.; Yan, X. Accelerated reaction kinetics in microdroplets: Overview and recent developments. *Annu. Rev. Phys. Chem.* **2020**, *71*, 31–51. [CrossRef] [PubMed]
3. Banerjee, S.; Zare, R.N. Syntheses of isoquinoline and substituted quinolines in charged microdroplets. *Angew. Chem. Int. Ed. Engl.* **2015**, *127*, 15008–15012. [CrossRef]

4. Espy, R.D.; Wleklinski, M.; Yan, X.; Cooks, R.G. Beyond the flask: Reactions on the fly in ambient mass spectrometry. *TrAC Trends Anal. Chem.* **2014**, *57*, 135–146. [CrossRef]
5. Yan, X.; Bain, R.M.; Cooks, R.G. Organic reactions in microdroplets: Reaction acceleration revealed by mass spectrometry. *Angew. Chem. Int. Ed. Engl.* **2016**, *55*, 12960–12972. [CrossRef]
6. Badu-Tawiah, A.K.; Cyriac, J.; Cooks, R.G. Reactions of organic ions at ambient surfaces in a solvent-free environment. *J. Am. Soc. Mass Spectrom.* **2012**, *23*, 842–849. [CrossRef]
7. Ansu-Gyeabourh, E.; Amoah, E.; Ganesa, C.; Badu-Tawiah, A.K. Monoacylation of symmetrical diamines in charge microdroplets. *J. Am. Soc. Mass Spectrom.* **2020**, *32*, 531–536. [CrossRef]
8. Huang, K.-H.; Wei, Z.; Cooks, R.G. Accelerated reactions of amines with carbon dioxide driven by superacid at the microdroplet interface. *Chem. Sci.* **2021**, *12*, 2242–2250. [CrossRef]
9. Zhao, P.; Gunawardena, H.P.; Zhong, X.; Zare, R.N.; Chen, H. Microdroplet ultrafast reactions speed antibody characterization. *Anal. Chem.* **2021**, *93*, 3997–4005. [CrossRef]
10. Kang, J.; Lhee, S.; Lee, J.K.; Zare, R.N.; Nam, H.G. Restricted intramolecular rotation of fluorescent molecular rotors at the periphery of aqueous microdroplets in oil. *Sci. Rep.* **2020**, *10*, 16859. [CrossRef]
11. Zhong, X.; Chen, H.; Zare, R.N. Ultrafast enzymatic digestion of proteins by microdroplet mass spectrometry. *Nat. Commun.* **2020**, *11*, 1049. [CrossRef] [PubMed]
12. Nam, I.; Lee, J.K.; Nam, H.G.; Zare, R.N. Abiotic production of sugar phosphates and uridine ribonucleoside in aqueous microdroplets. *Proc. Natl. Acad. Sci. USA* **2017**, *114*, 12396–12400. [CrossRef] [PubMed]
13. Qiu, L.; Cooks, R.G. Simultaneous and spontaneous oxidation and reduction in microdroplets by the water radical cation/anion pair. *Angew. Chem. Int. Ed. Engl.* **2022**, *134*, e202210765. [CrossRef]
14. Holden, D.T.; Morato, N.M.; Cooks, R.G. Aqueous microdroplets enable abiotic synthesis and chain extension of unique peptide isomers from free amino acids. *Proc. Natl. Acad. Sci. USA* **2022**, *119*, e2212642119. [CrossRef] [PubMed]
15. Cheng, H.; Tang, S.; Yang, T.; Xu, S.; Yan, X. Accelerating electrochemical reactions in a voltage-controlled interfacial microreactor. *Angew. Chem. Int. Ed. Engl.* **2020**, *59*, 19862–19867. [CrossRef] [PubMed]
16. Müller, T.; Badu-Tawiah, A.; Cooks, R.G. Accelerated Carbon Carbon Bond-Forming Reactions in Preparative Electrospray. *Angew. Chem. Int. Ed. Engl.* **2012**, *124*, 12002–12005. [CrossRef]
17. He, Q.; Badu-Tawiah, A.K.; Chen, S.; Xiong, C.; Liu, H.; Zhou, Y.; Hou, J.; Zhang, N.; Li, Y.; Xie, X. In Situ bioconjugation and ambient surface modification using reactive charged droplets. *Anal. Chem.* **2015**, *87*, 3144–3148. [CrossRef]
18. Wei, Z.; Wleklinski, M.; Ferreira, C.; Cooks, R.G. Reaction acceleration in thin films with continuous product deposition for organic synthesis. *Angew. Chem. Int. Ed. Engl.* **2017**, *129*, 9514–9518. [CrossRef]
19. Tata, A.; Salvitti, C.; Pepi, F. From vacuum to atmospheric pressure: A review of ambient ion soft landing. *J. Am. Soc. Mass Spectrom.* **2020**, *450*, 116309. [CrossRef]
20. Wei, Z.; Zhang, X.; Wang, J.; Zhang, S.; Zhang, X.; Cooks, R.G. High yield accelerated reactions in nonvolatile microthin films: Chemical derivatization for analysis of single-cell intracellular fluid. *Chem. Sci.* **2018**, *9*, 7779–7786. [CrossRef]
21. Salvitti, C.; de Petris, G.; Troiani, A.; Managò, M.; Villani, C.; Ciogli, A.; Sorato, A.; Ricci, A.; Pepi, F. Accelerated D- Fructose Acid-Catalyzed Reactions in Thin Films Formed by Charged Microdroplets Deposition. *J. Am. Soc. Mass Spectrom.* **2022**, *33*, 565–572. [CrossRef] [PubMed]
22. Troiani, A.; de Petris, G.; Pepi, F.; Garzoli, S.; Salvitti, C.; Rosi, M.; Ricci, A. Base-assisted conversion of protonated D-fructose to 5-HMF: Searching for gas-phase green models. *Chem. Open* **2019**, *8*, 1190. [CrossRef] [PubMed]
23. Salvitti, C.; de Petris, G.; Troiani, A.; Managò, M.; Ricci, A.; Pepi, F. Kinetic Study of the Maillard Reaction in Thin Film Generated by Microdroplets Deposition. *Molecules* **2022**, *27*, 5747. [CrossRef]
24. Salvitti, C.; de Petris, G.; Troiani, A.; Managò, M.; Di Noi, A.; Ricci, A.; Pepi, F. Sulfuric Acid Catalyzed Esterification of Amino Acids in Thin Film. *J. Am. Soc. Mass Spectrom.* **2023**, *34*, 2748–2754. [CrossRef] [PubMed]
25. De Angelis, M.; Managò, M.; Pepi, F.; Salvitti, C.; Troiani, A.; Villani, C.; Ciogli, A. Stereoselectivity in electrosprayed confined volumes: Asymmetric synthesis of warfarin by diamine organocatalysts in microdroplets and thin films. *RSC Adv.* **2024**, *14*, 1576–1580. [CrossRef] [PubMed]
26. Li, A.; Luo, Q.; Park, S.; Cooks, R.G. Synthesis and catalytic reactions of nanoparticles formed by electrospray ionization of coniage metals. *Angew. Chem. Int. Ed. Engl.* **2014**, *53*, 3147–3150. [CrossRef]
27. Li, A.; Baird, Z.; Bag, S.; Sarkar, D.; Prabhath, A.; Pradeep, T.; Cooks, R.G. Using ambient ion beams to write nanostructured patterns for Surface Enhanced Raman spectroscopy. *Angew. Chem. Int. Ed. Engl.* **2014**, *53*, 12528–12531. [CrossRef]
28. Basuri, P.; Chakraborty, A.; Ahuja, T.; Mondal, B.; Kumar, J.S.; Pradeep, T. Spatial reorganization of analytes in charged aqueous microdroplets. *Chem. Sci.* **2022**, *13*, 13321–13329. [CrossRef]
29. Jin, S.; Chen, H.; Yuan, X.; Xing, D.; Wang, R.; Zhao, L.; Zhang, D.; Gong, C.; Zhu, C.; Gao, X. The Spontaneous electron-mediated redox processes on sprayed water microdroplets. *JACS Au* **2023**, *3*, 1563–1571. [CrossRef]
30. Kafeenah, H.; Jen, H.; Chen, S. Microdroplet mass spectrometry: Accelerating reaction and application. *Electrophoresis* **2022**, *43*, 74–81. [CrossRef]
31. Sarfraz, N.; Khan, I. Plasmonic gold nanoparticles (AuNPs): Properties, synthesis and their advanced energy, environmental and biomedical applications. *Chem. Asian J.* **2021**, *16*, 720–742. [CrossRef] [PubMed]

32. Lee, J.K.; Samanta, D.; Nam, H.G.; Zare, R.N. Spontaneous formation of gold nanostructures in aqueous microdroplets. *Nat Commun.* **2018**, *9*, 1562. [CrossRef] [PubMed]
33. Salvitti, C.; Troiani, A.; Mazzei, F.; D'Agostino, C.; Zumpano, R.; Baldacchini, C.; Bizzarri, A.R.; Tata, A.; Pepi, F. The use of a commercial ESI Z-spray source for ambient ion soft landing and microdroplet reactivity experiments. *Int. J. Mass Spectrom.* **2021**, *468*, 116658. [CrossRef]
34. Badu-Tawiah, A.K.; Campbell, D.I.; Cooks, R.G. Reactions of microsolvated organic compounds at ambient surfaces: Droplet velocity, charge state, and solvent effects. *J. Am. Soc. Mass Spectrom.* **2012**, *23*, 1077–1084. [CrossRef]
35. Mondal, S.; Acharya, S.; Biswas, R.; Bagchi, B.; Zare, R.N. Enhancement of reaction rate in small-sized droplets: A combined analytical and simulation study. *J. Chem. Phys.* **2018**, *148*, 244704. [CrossRef]
36. Zumpano, R.; Polli, F.; D'Agostino, C.; Antiochia, R.; Favero, G.; Mazzei, F. Nanostructure-Based Electrochemical Immunosensors as Diagnostic Tools. *Electrochem* **2021**, *2*, 10–28. [CrossRef]
37. Lazanas, A.C.; Prodromidis, M.I. Electrochemical Impedance Spectroscopy—A Tutorial. *ACS Meas. Sci. Au* **2023**, *3*, 162–193. [CrossRef]
38. Choi, W.; Shin, H.-C.; Kim, J.M.; Choi, J.-Y.; Yoon, W.-S. Modeling and Applications of Eletrochemical Impedance Spectroscopy (EIS) for Lithium-ion Batteries. *J. Electrochem. Sci. Technol.* **2020**, *11*, 1–13. [CrossRef]
39. Loveday, D.; Peterson, P.; Rodgers, B. Evaluation of organic coatings with electrochemical impedance spectroscopy. *JCT Coat. Tech* **2004**, *8*, 46–52.
40. Ariyoshi, K.; Siroma, Z.; Mineshige, A.; Takeno, M.; Fukutsuka, T.; Abe, T.; Uchida, S. Electrochemical Impedance Spectroscopy Part 1: Fundamentals. *Electrochemistry* **2022**, *90*, 102007. [CrossRef]
41. Pumera, M.; Aldavert, M.; Mills, C.; Merkoçi, A.; Alegret, S. Direct voltammetric determination of gold nanoparticles using graphite-epoxy composite electrode. *Electrochim. Acta* **2005**, *50*, 3702–3707. [CrossRef]
42. Edgeccombe, C.J.; Valdre, U. Microscopy and computational modelling to elucidate the enhancement factor for field electron emitters. *J. Microsc.* **2001**, *203*, 188–194. [CrossRef]
43. Stassi, S.; Cauda, V.; Canavese, G.; Manfredi, D.; Pirri, C.F. Synthesis and characterization of gold nanostars as filler of tunneling conductive polymer composites. *Eur. J. Inorg. Chem.* **2012**, *16*, 2669–2673. [CrossRef]
44. Göver, T.; Zafer, Y. Electrochemical study of 6-(ferrocenyl) hexanethiol on gold electrode surface in non-aqueous media. *Surf. Interfaces* **2018**, *13*, 163–167. [CrossRef]
45. Wang, B.; Anzai, J.I. A facile electrochemical detection of hypochlorite ion based on ferrocene compounds. *Int. J. Electrochem. Sci.* **2015**, *10*, 3260–3268. [CrossRef]
46. Zhang, H.; Hu, H.; Li, Y.; Wang, J.; Ma, L. A ferrocene-based hydrogel as flexible electrochemical biosensor for oxidative stress detection and antioxidation treatment. *Biosens. Bioelectron.* **2024**, *248*, 115997. [CrossRef]

surfaces

MDPI

Article

Kelvin Probe Force Microscopy, Current Mapping, and Optical Properties of Hybrid ZnO Nanorods/Ag Nanoparticles

Ishaq Musa

Department of Physics, Palestine Technical University-Kadoorie, Tulkarm P.O. Box 7, Palestine; i.musa@ptuk.edu.ps

Abstract: The optical characteristics and electrical behavior of zinc oxide nanorods (ZnO-NRs) and silver nanoparticles (Ag-NPs) were investigated using advanced scanning probe microscopy techniques. The study revealed that the ZnO nanorods had a length of about 350 nm, while the Ag nanoparticles were spherical with heights ranging from 5 to 14 nm. Measurements with Kelvin probe force microscopy (KPFM) showed that the work functions of ZnO nanorods were approximately 4.55 eV, higher than that of bulk ZnO, and the work function of Ag nanoparticles ranged from 4.54 to 4.56 eV. The electrical characterization of ZnO nanorods, silver nanoparticles, and their hybrid was also conducted using conductive atomic force microscopy (C-AFM) to determine the local current-voltage (I-V) characteristics, which revealed a characteristic similar to that of a Schottky diode. The current-voltage characteristic curves of ZnO nanorods and Ag nanoparticles both showed an increase in current at around 1 V, and the hybrid ZnONRs/AgNP exhibited an increase in turn-on voltage at around 2.5 volts. This suggested that the presence of Ag nanoparticles enhanced the electrical properties of ZnO nanorods, improving the charge carrier mobility and conduction mechanisms through a Schottky junction. The investigation also explored the optical properties of ZnO-NRs, AgNPs, and their hybrid, revealing absorption bands at 3.11 eV and 3.18 eV for ZnO-NRs and AgNPs, respectively. The hybrid material showed absorption at 3.13 eV, indicating enhanced absorption, and the presence of AgNP affected the optical properties of ZnO-NR, resulting in increased photoluminescence intensity and slightly changes in peak positions.

Keywords: KPFM; ZnO-NR; Ag-NP; Schottky diode; absorption; photoluminescence

1. Introduction

The intrinsic properties of ZnO, such as its wide bandgap energy, high melting point, thermal stability, strong piezoelectric effect, hexagonal wurtzite structure, and large exciton binding energy, have established it as a popular material in wide bandgap semiconductors [1,2]. These properties have led to its utilization in optical and optoelectronic applications, piezoelectric nanogenerators, gas sensing, and multifunctional nanodevices. Nevertheless, research has indicated that the closure of intrinsic defects and the shape-dependent properties of nanorods can significantly improve their performance. This positions ZnO as an ideal semiconductor oxide material for emerging applications and a model material for examining the properties of low-dimensional nanostructures [3–5].

Specifically, because of ZnO's potential practical uses and numerous reports on matrix-doped ZnO, one of the most promising options is when ZnO is doped with silver nanoparticles. This addition has been proven to significantly improve the electrical and optical properties of ZnO nanorods, making them a desirable choice for various technological advancements in the fields of electronics, photonics, and optoelectronics [6,7].

Silver nanoparticles (AgNPs) have exceptional optical and electrical properties that have generated significant interest in scientific and technological fields. These unique characteristics are due to their small size and quantum mechanical effects. Optically, silver nanoparticles are recognized for their intense surface plasmon resonance (SPR), where conduction electrons on the nanoparticle surface oscillate in response to light, leading to strong absorption and scattering of specific wavelengths and resulting in vibrant colors and enhanced electromagnetic fields near the particle surface. Electrically, silver nanoparticles have excellent conductivity due to their high electron mobility and large surface area-to-volume ratio, making them ideal for use in electronics, conductive inks, and sensors. Additionally, the quantum confinement effects in nanoparticles influence their electronic properties, leading to discrete energy levels and potentially enhanced reactivity compared with bulk silver [8–11].

To gain a deeper understanding and improve the properties of nanostructured materials, it is crucial to utilize advanced microscopy techniques. Popular techniques include techniques such as Kelvin probe force microscopy (KPFM) and current mapping properties at the nanometer scale. These methods produce clearer potential images and are essential for determining the electrical properties of nanostructured materials, thus aiding in the development of advanced nanotechnologies [12–14].

KPFM utilizes a dual-pass technique where a modulated voltage is applied to the cantilever tip as the oscillation frequency, followed by a phase-sensitive detection system to capture the frequency shift and the tip's work function relative to the sample. The alteration in the tip's work function induces an AC bias potential difference between the tip and the sample, moving toward Fermi level equalization, deduced by the thermionic field emission effect. The rectified incremental bias ΔV is related to the work function difference between the tip and the sample. This allows KPFM to study work functions on surfaces at a high resolution, down to the nanometer scale. KPFM is versatile, accurate, and non-destructive, making it suitable for studying electronic materials for next-generation electronic devices [15–19].

Electricity characteristics of materials on a nanoscale can be studied using current mapping and current-voltage (I-V) measurements with scanning probe microscopy (SPM). In current mapping, a conductive probe moves across the material's surface in constant contact or tapping mode, detecting local variations in electrical conductivity. This method offers precise spatial resolution, revealing different conductive areas and their electrical behaviors [20]. On the other hand, I-V measurements involve applying a variable voltage between the probe and the sample while recording the resulting current. This helps determine important electronic properties like resistance, diode behavior, and charge carrier mobility. When used together, current mapping and I-V measurements reveal detailed information about electronic heterogeneity, charge transport mechanisms, and the impact of nanostructures on material performance [21–24].

The information gathered from the samples studied, including their work function, surface potential, current mapping, and band structure properties, allows us to gain a deeper understanding of their physical origins.

The primary focus of this research was to understand the fundamental properties of hybrid ZnO nanorods and silver nanoparticles. Electrical and optical characterization were performed to verify the effectiveness of the hybrid system.

This study aimed to comprehensively explore the work function and electrostatic force microscopy of zinc oxide (ZnO) nanorods and silver (Ag) nanoparticles. In addition to that, we thoroughly examined the optical properties of the hybrid nanostructure of ZnO nanorods and silver nanoparticles. Moreover, we conducted extensive investigations into the current-voltage characteristics of both ZnO nanorods and Ag nanoparticles, as well as the dynamic behavior of the hybrid structure, utilizing state-of-the-art current mapping techniques.

2. Materials and Methods

2.1. ZnO Nanorods and Ag Nanoparticles Synthesis

In our previous research [25], we successfully generated ZnO nanorods by first heating a solution containing 5.5 g of zinc acetate dehydrate (98+%, Sigma Aldrich, Taufkirchen, Germany) in 250 mL of ethanol until it became clear. The solution was then refluxed for

1 h, with 150 mL of the solvent removed by distillation and replaced with fresh ethanol. Subsequently, 1.39 g of lithium hydroxide monohydrate (Aldrich) was added to the solution in an ultrasonic bath at 0 °C and dispersed for 1 h, resulting in a transparent solution of ZnO sol-gel nanoparticles. To create the ZnO nanorods, the solution containing ZnO nanoparticles was heated and combined with 10% distilled water (DW) at 60 °C for 48 h, resulting in the formation of a white powder precipitate.

In another recent work [26], silver nanoparticles were synthesized by mixing 1.18 mM $AgNO_3$ aqueous solutions with deionized water. Subsequently, 50 mL of *Pistacia palaestina* (*P. palaestina*) leaf extract was added drop by drop to each 50 mL $AgNO_3$ solution. The mixture was then warmed in a heating mantle, maintaining a temperature range of 80 to 84 °C for 2 h with continuous stirring. The successful formation of silver nanoparticles was indicated by the emergence of a brownish-yellow to black color. To purify the synthesized nanoparticles, the solution was centrifuged at 10,000 rpm for 10 min, a process that was repeated five times to ensure the retrieval of pure silver nanoparticles. Afterward, the hybrid composite of AgNP/ZnONR was formed by adding 10% of AgNP to ZnONR and then exposing it to sonication for 30 min.

2.2. Deposition and Characterization Methods

We deposited ZnO nanorods and silver nanoparticles on various substrates such as mica sheet, P-type silicon substrate, metal steel, and glass for different characterization purposes. The morphology, KPFM, and current mapping characteristics of the ZnO nanorods, silver nanoparticles, and their hybrid were evaluated using scanning probe microscopy (SPM-9700HT, Shimadzu, Tokyo, Japan). To analyze the electrical conductivity and work function, we utilized AFM and KPFM for these measurements. These methodologies allowed for precise determinations of the nanoparticles' conductivity, morphology, and surface potential. The samples were carefully placed on a piezoelectric stage for examination, and a conductive probe made of PtSi was attached to the end of the cantilever. This setup allowed for a uniform oscillatory motion as the probe traversed the sample's surface, providing a comprehensive map of its topography. KPFM further enriched our analysis by generating images of the surface potential alongside the topographic map at each point of the scan. The interaction between the probe and the sample surface included an electrostatic force component, originating from the contact potential difference (CPD) between them, providing valuable data on the surface potential. Proper preparation of the samples was crucial for precise topographic and potential mapping, with each sample fixed to carbon tape and securely fastened to a steel disc to ensure stability during scanning. The scanning was performed at a speed of 0.5 Hz and a resolution of 256 × 256 pixels, delivering high-definition images, and the instrument's spatial resolution was recorded at 0.2 nm. Localized nanoscale current mapping measurements were conducted using the C-AFM (conductive atomic force microscopy) technique, which is widely recognized for its high spatial resolution. To obtain the measurements, Nanoworld supplied Pt/Ir-coated tips with a resonance frequency of 13 kHz and a force constant of 0.2 N/m, selected for their excellent conductive properties. These tips were used as the top electrode for the purpose of understanding the current-voltage (I-V) characteristics using the C-AFM technique. The I-V response, which offered valuable insight into the electrical behavior of the sample, was carefully derived from the recorded current images. This was achieved by sweeping the bias voltage within the range of −5 V to +5 V while keeping a constant sample bias of 3 V. The X-ray diffraction analysis was conducted using a Bruker D2 PHASER (Bruker, Billerica, MA, USA) with Cu kα radiation settings of 30 kV and 10 mA. The UV–vis absorption spectra were acquired using a UV-2600i spectrophotometer (Shimadzu, Tokyo, Japan), and the photoluminescence spectra were captured with an RF-6000 spectrofluorometer (Shimadzu, Tokyo, Japan). Data collection was performed using Shimadzu LabSolutions UV-Vis software.

3. Results and Discussion

3.1. Morphology and Structure of ZnO-NRs and Ag-NPs

The shape, length, and height of the ZnO nanorods and Ag nanoparticles were carefully examined using scanning probe microscopy (SPM) topography images captured in a non-contact dynamic mode. In order to conduct this examination, a Super Sharp Silicon (SSS-NCH) AFM tip from Nanoworld, possessing a force constant of 42 N/m, a resonance frequency of 320 kHz, and a tip radius of 2 nm, was utilized. As depicted in Figure 1a–c, the provided atomic force microscopy (AFM) topographic representation of ZnO nanorods on a Si substrate successfully revealed the morphology and length of the nanorods, which were determined to be approximately 350 nm, as illustrated in Figure 1c. Moving on to Figure 1d–f, it displays silver nanoparticles on a mica substrate, showcasing their spherical morphology quite explicitly. Conducting a height analysis as shown in Figure 1f facilitated the determination that the observed heights of these nanoparticles ranged from 5 to 14 nm. Finally, Figure 1g–i showcases hybrid ZnO nanorods and silver nanoparticles.

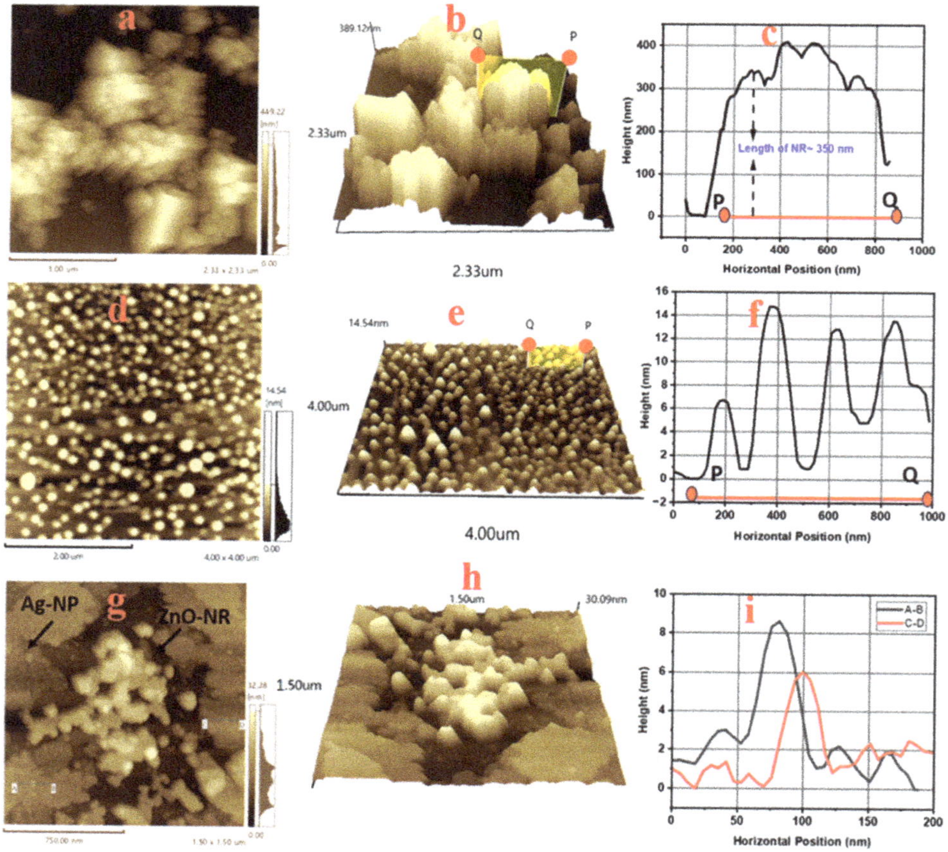

Figure 1. (**a**) Atomic force microscopy (AFM) topography images of ZnO nanorods on a Si substrate. Scan sizes: (2.33 μm × 2.33 μm). (**b**) Three-dimensional projection topography of ZnO-NR. (**c**) Line profiles for ZnO-NRs identified in the image in (**b**). (**d**) AFM topography image of Ag nanoparticles on a mica substrate. Scan sizes: (4 μm × 4 μm). (**e**) Three-dimensional projection topography of Ag-NP. (**f**) Line profiles for AgNP identified in the image in (**e**). (**g**) AFM topography image of hybrid Ag NP and ZnO NR on a Si substrate. Scan sizes: (1.5 μm × 1.5 μm). (**h**) Three-dimensional projection topography of hybrid Ag NP and ZnO-NR. (**i**) Line profiles for AgNP identified in the image in (**h**).

For more details about the crystal structure of the ZnO nanorods and Ag nanoparticles, see Figure 2, which illustrates a typical XRD pattern of the ZnO nanorods. The sharpness of the peaks signified the high level of crystallinity of the ZnO nanorods, demonstrating a single-phase wurtzite structure. The observed diffraction peaks were associated with the existence of a pure hexagonal phase of ZnO with lattice constants of a = 3.251 Å and c = 5.208 Å, consistent with JCPDS card for ZnO (JCPDS, 65-3411) [27]. The crystalline diameter of ZnO nanorods was approximated to be 20 nm, calculated employing the Debye–Scherer formula. This formula takes into account various factors, including the crystallite size, wavelength of the X-ray used, and the full width at half maximum (FWHM) of the observed diffraction peak. By applying this formula, we can estimate the average crystalline size of the ZnO nanorods, providing crucial information about their physical properties [28]. Furthermore, the XRD patterns of silver nanoparticles revealed peaks at 2θ angles of 38.4°, 44.7°, 64.6°, and 77.7°, corresponding to the (111), (200), (220), and (311) planes, indicating the crystalline phase of silver metal with a face-centered cubic structure, consistent with the JCPDS File No. 04-0783. These diffraction peaks allowed us to determine the orientation and arrangement of the atoms within the silver nanoparticles, providing valuable insights into their structural properties [29]. The average size of Ag nanoparticles, calculated from the full width at half maximum (FWHM) of the peak for the 111 plane, was 14.8 nm. The micro-strain (ε) for ZnO nanorods was 0.0019, while the micro-strain (ε) for AgNP was 0.0023. The micro-strain could be determined using the equation $\varepsilon = \beta\cos\theta/4$, where β is the line broadening and θ is the diffraction angle [28].

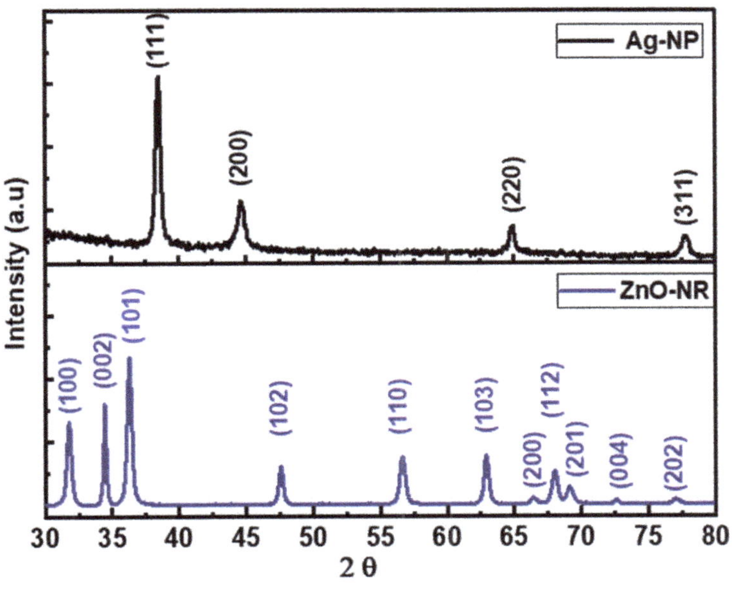

Figure 2. The XRD pattern of ZnO nanorods and Ag nanoparticles.

3.2. Kelvin Probe Force Microscopy of ZnO-NRs and AgNPs

Figures 3 and 4 show the KPFM characteristics of ZnO nanorods and Ag nanoparticles on a silicon substrate. The detailed images reveal the electrical interaction between the probe and the sample, with high potential shown as red and low potential as yellow for the ZnO nanorods and Ag nanoparticles. The average contact potential difference (CPD) for ZnO nanorods with a 30 nm diameter and 400 nm length was approximately 96 mV. Similarly, ZnO nanorods with a 20 nm diameter and 220 nm lengths had a contact potential difference of 81 mV. In contrast, nanorods with an 8 nm diameter and 130 nm length showed a contact potential difference of 56 mV. These results emphasized the influence of

ZnO nanorod dimensions on the contact potential difference, as illustrated in Figure 3c,d. The AFM topography and KPFM images of Ag nanoparticles of varying sizes are shown in Figure 4. The average CPD for the Ag nanoparticles was 46 mV for a size of 4 nm and 70 mV for a size of 13 nm, as illustrated in Figure 4d. Figure 5 displays the KPFM for a reference sample of highly ordered pyrolytic graphite (HOPG), grade ZYA. This sample had a mosaic spread of $0.4 \pm 0.1^{\circ}$ and dimensions of 10 mm by 10 mm with a thickness of 1 mm. The average contact potential difference (CPD) for this sample was approximately 58 mV, as shown in Figure 5b. The sample was provided by (MikroMasch, Wetzlar, Germany).

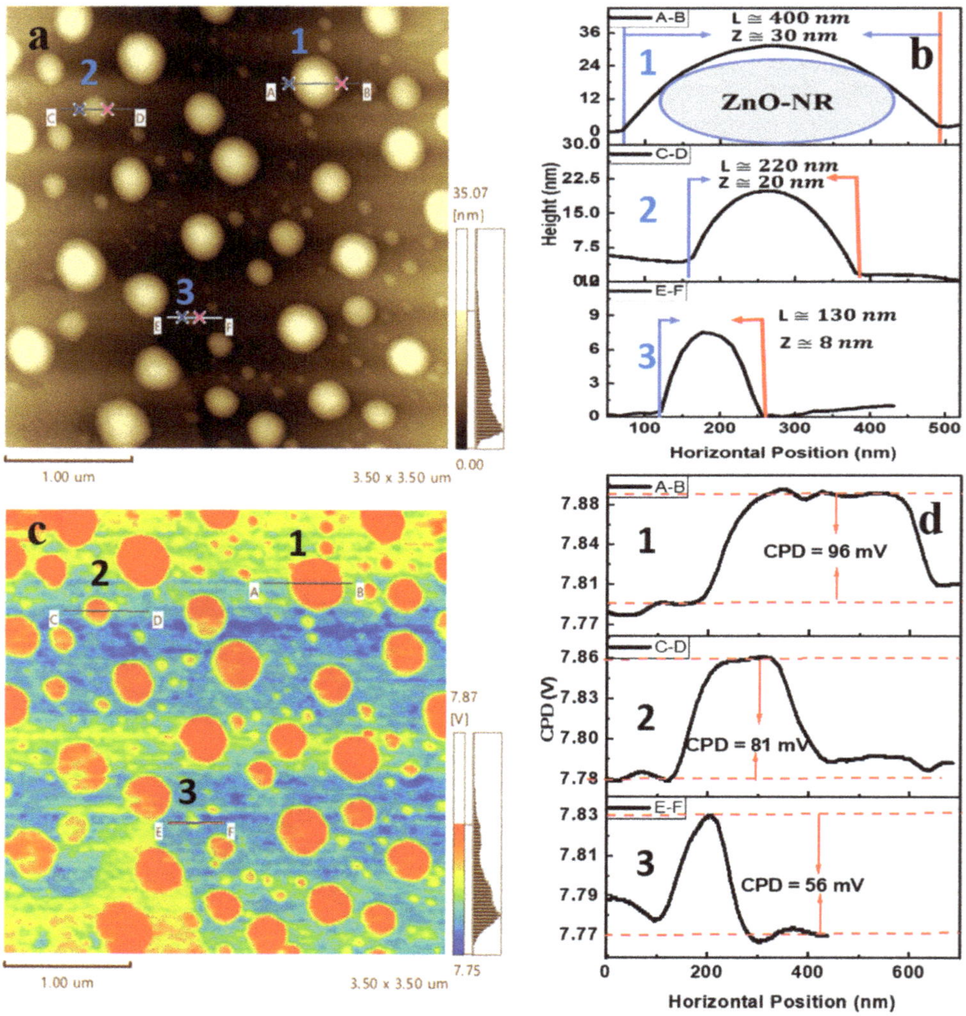

Figure 3. (**a**) AFM topography images of ZnO nanorods on a Si substrate. Scan sizes: (3.5 μm × 3.5 μm). (**b**) Line profiles for variable size and length of ZnO-NRs identified in the image in (**a**). (**c**) Surface potential images of ZnO NRs deposited on the silicon substrate. (**d**) CPD line profiles for different sizes and lengths of ZnO NRs identified in the image in (**c**).

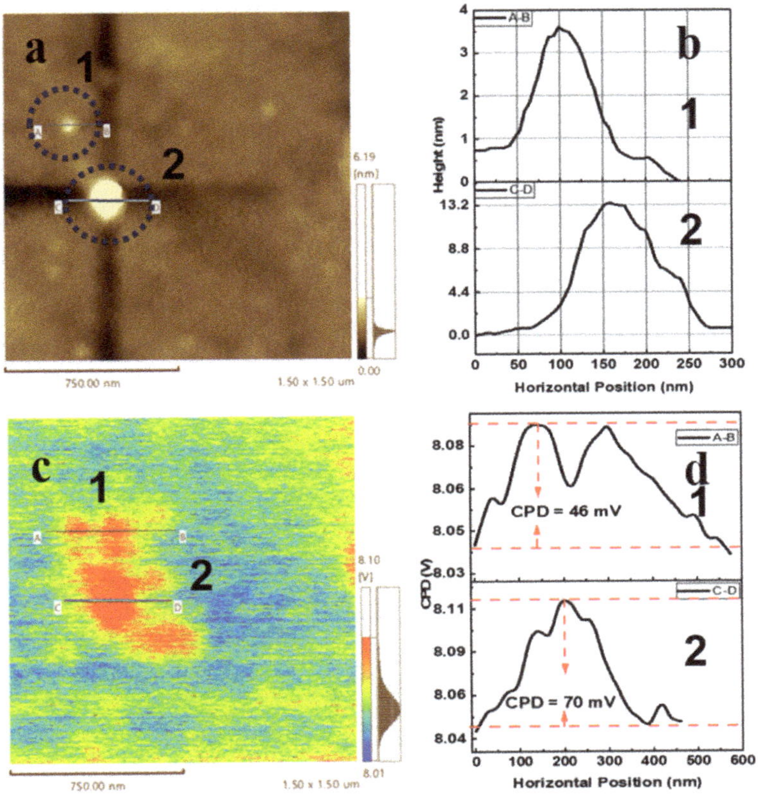

Figure 4. (**a**) AFM topography images of Ag nanoparticles on a Si substrate. Scan sizes: (3.5 μm × 3.5 μm). (**b**) Line profiles for variable Ag nanoparticles identified in the image in (**a**). (**c**) Surface potential images of Ag nanoparticles deposited on the silicon substrate. (**d**) CPD line profiles for different sizes of Ag nanoparticles identified in the image in (**c**).

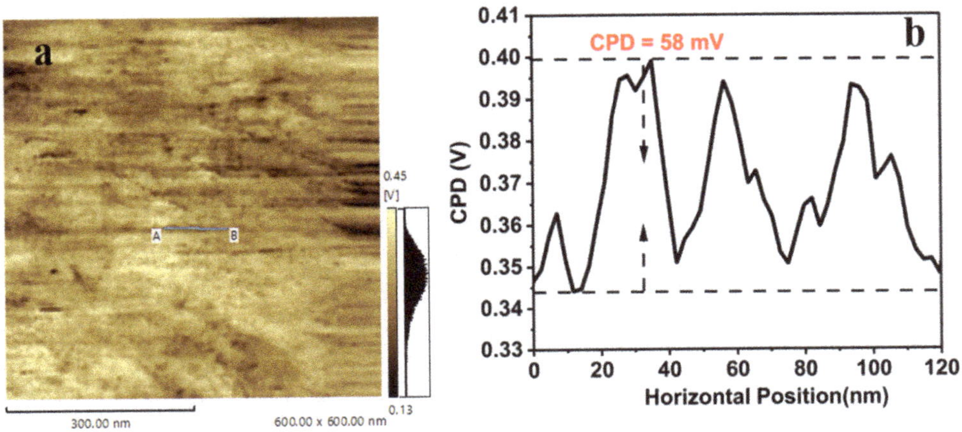

Figure 5. (**a**) Surface potential of reference sample highly ordered pyrolytic graphite (HOPG) and (**b**) height profile (CPD) of HOPG extract from image (**a**).

The main goal of Kelvin probe force microscopy (KPFM) is to accurately measure the work function of a specific sample. A key factor in KPFM is the surface potential, which represents the difference in work function values between the sample surface, denoted as ϕ_{sample}, and the tip of the AFM probe, represented as ϕ_{tip}. This difference is mathematically defined by the following equation:

$$\phi_{sample} = \phi_{tip} - eV_{CPD} \quad (1)$$

where e represents the charge of a single electron, and V_{CPD} is the contact potential difference between the probe tip and the sample's surface [30]. To obtain an accurate assessment of the work function of ZnO nanorods, it is necessary to determine the work function of the cantilever within the apparatus. Calibration is important to ensure the reliability of measurements and typically involves the use of a reference material with a known work function for standardization purposes. In this study, highly oriented pyrolytic graphite (HOPG) was used as the reference material, with a work function range of 4.5 to 5 eV, serving as a benchmark for calibration and precision of measurements [31].

Using Equation (1), similar formulas are derived for both ZnO nanorods and HOPG substrates. The work function of the ZnO nanorods substrate is denoted as ϕ_{ZnO-NR}, and the specific contact potential difference for ZnO nanorods substrate is represented as $V_{(CPD,ZnO\text{-}NR)}$. Similarly, the work function for the HOPG substrate is ϕ_{HOPG}, and its unique contact potential difference is $V_{(CPD,HOPG)}$. The difference between the work functions of these two substrates can be determined using the following equations:

$$\phi_{ZnO-NR} = \phi_{HOPG} + e\left(V_{(CPD,HOPG)} - V_{(CPD,ZnO-NR)}\right) \quad (2)$$

Using the Equation (2) and CPD values from the analysis in Figures 3c,d and 5b and the published values for HOPG's work function, the work function of ZnO nanorods was accurately estimated. For ZnO nanorods with approximately 30 nm diameter and 400 nm length, the work function was calculated to be 4.556 eV. Additionally, for ZnO nanorods with a diameter of around 20 nm and a length of 220 nm, the work function was estimated to be 4.527 eV. On the other hand, for ZnO nanorods with a diameter of 8 nm and a length of 130 nm, the work function was found to be 4.552 eV. Our calculation of the work function for ZnO nanorods was approximately 4.55 eV, which was higher than the work function of ZnO bulk, typically around 4.45 eV [32]. The difference in work function values observed between ZnO nanorods and bulk ZnO was attributed to various factors. Quantum confinement, influenced by the nanorods' size, restricted electron movement, leading to higher energy levels and an increased work function. Nanorods also had a greater surface-to-volume ratio compared with bulk materials, leading to changes in electronic structure and possibly increased surface energy, resulting in a higher work function [33].

In the context of calculating the work function of silver nanoparticles, we use Equation (1) and can also create similar equations for both Ag nanoparticles and HOPG substrates. By representing the work function of the silver nanocluster substrate as ϕ_{Ag} and the specific contact potential difference for the silver nanoparticles substrate as $V_{(CPD,Ag)}$, we can then denote the work function of the HOPG substrate as ϕ_{HOPG} and its unique contact potential difference as $V_{(CPD,HOPG)}$. The relationship between the work functions of silver and HOPG is connected through their respective contact potential differences:

$$\phi_{Ag} = \phi_{HOPG} + e\left(V_{(CPD,HOPG)} - V_{(CPD,Ag)}\right) \quad (3)$$

Using Equation (3) and the contact potential difference (CPD) data from Figures 4d and 5b, along with the work function value of HOPG from the literature [32], the work function of the AgNPs with an average diameter of 4 nm was determined to be 4.56 eV. For nanoparticles of size 14 nm, the work function equaled 4.538 eV.

From the literature sources, the work function of bulk silver was known to be 4.3 eV, while the work function of the Ag nanocluster was 4.587 eV [34], and that of a vacuum-deposited Ag electrode was 4.68 eV. The work function of the Ag nanoparticles was estimated to be around 4.56 eV and 4.54, which aligned closely with the values reported in the literature. The remarkable findings revealed that an increase in particle size led to a significant decrease in work function due to the interesting isotropic surface curvature of the symmetrical spherical shape. Equation (4) is utilized to compute the electronic work function for a spherical particle, taking into account the work function for the plane surface, electron charge, and the permittivity of the medium. This equation emphasizes that the surface characteristics of a nanoparticle have a greater impact when the particle is smaller [35,36].

$$\phi_E^s = \phi_E^0 + \frac{1}{2\pi\epsilon}\left(\frac{e^2}{r_a}\right) \tag{4}$$

The constant curvature of a spherical nanoparticle greatly increases its work function compared with a flat surface, requiring more energy to extract an electron from a surface with a higher degree of positive curvature. This compelling example effectively demonstrates the complex interaction between quantum mechanics and electrostatics on the mesmerizing nanoscale, ultimately leading to the emergence of size-specific properties distinct from those observed in bulk materials.

3.3. *Current Mapping and I-V Measurements*

We present an electrical analysis of zinc oxide (ZnO) nanorods, silver nanoparticles (AgNPs), and a hybrid material using conductive atomic force microscopy (C-AFM). This advanced method enabled the simultaneous recording of current maps and topography obtained through contact mode AFM. Additionally, C-AFM allowed for the precise determination of the local current-voltage (I-V) characteristics of the top surface of ZnO nanorods, silver nanoparticles, and the hybrid composite material as shown in Figures 6 and 7. By applying voltage through a conductive tip in contact with the ZnO nanorods' surfaces, variations in electrical current could be detected, unveiling defects and surface states. This method enabled precise control and adjustment of the electrical properties of the ZnO nanorods.

Distinct areas with heightened and reduced current values, depicted in Figure 6c, pointed out significant regions such as defects or highly conducting pathways. Scattered yet noteworthy regions of elevated current indicated localized areas of high conductivity or possible defects, with current values equal to or exceeding 150 nA. Conversely, widespread low-current sections covered most of the image, indicating general low conductivity or a consistent baseline current level throughout the ZnO nanorods, with current values equal to or less than 10 nA. Furthermore, for Ag nanoparticles, the high-current region ranged from 120 nA to 200 nA, possibly linked to differences in their size. Similarly, for hybrid ZnO-NR and Ag-NP, the high current region fell between 160 nA and 200 nA, as seen in Figure 6g.

The results presented in Figure 7 show a typical I-V characteristic curve obtained from a metallic disc as a reference sample, ZnO nanorods (ZnONRs), silver nanoparticles (Ag-NPs), and a combination of ZnO-NR and Ag-NP samples. The findings revealed a significant increase in current at approximately 1 V for both ZnO-NRs and Ag-NPs, indicating turn-on voltages of around 1 V for each. When combined, there was a different turn-on voltage around 2.5 V, suggesting a unique response to the combined effect. The reverse bias analysis also highlighted the distinctive characteristics of the ZnONR and Ag-NP combination, effectively blocking the current until very high negative voltages. It is important to understand that the presence of Ag-NP significantly impacted the electrical properties of ZnONR. The observed effects in the I-V traits for ZnO nanorods (ZnONRs), silver nanoparticles (Ag-NPs), and their combination required a thorough understanding of the physical and chemical interactions between these materials.

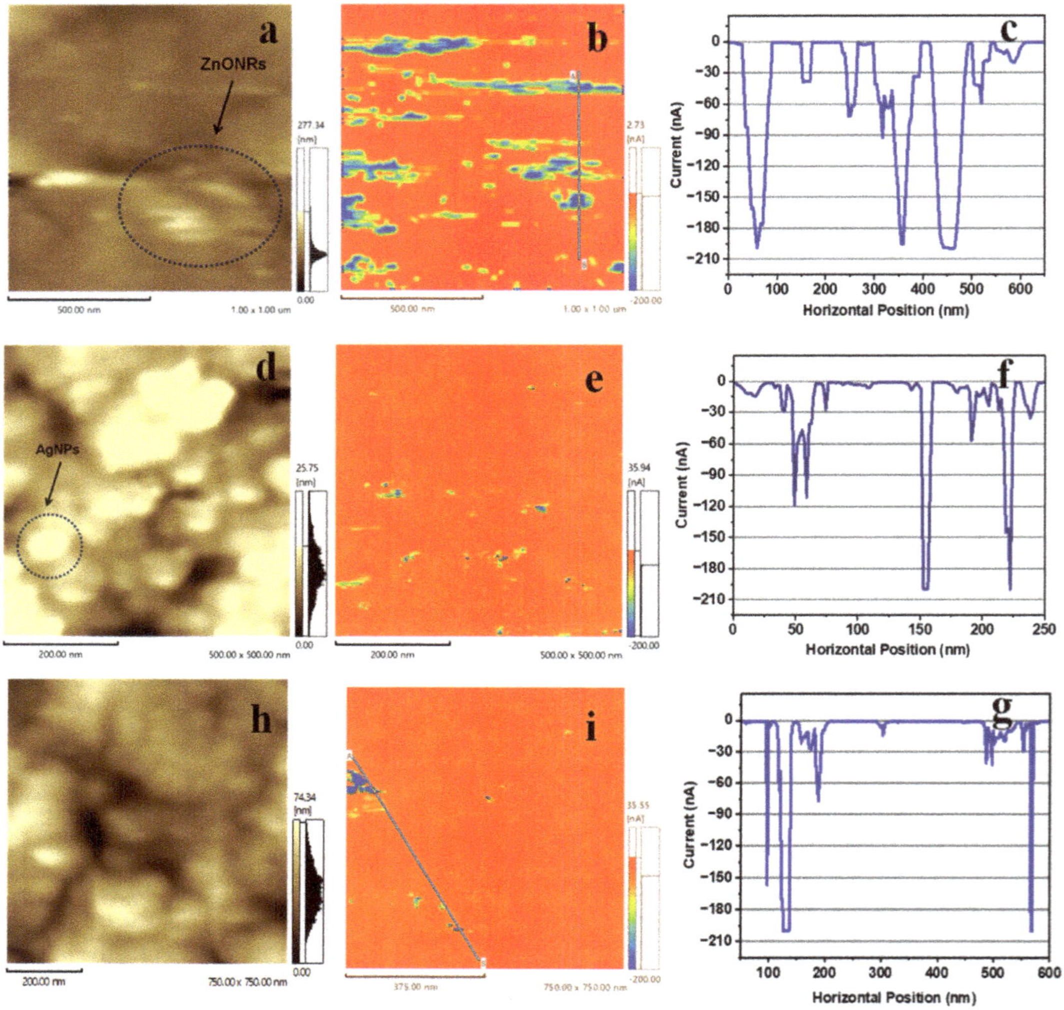

Figure 6. (**a**) AFM topography images of ZnO nanorods on a metallic disc substrate. Scan sizes: (1 µm × 1 µm). (**b**) Current mapping of ZnO-NR. (**c**) Line profiles of current for ZnO-NRs identified in the image in (**b**). (**d**) AFM topography image of Ag nanoparticles on a metallic disc substrate. Scan sizes: (0.5 µm × 0.5 µm). (**e**) Current mapping of Ag-NP. (**f**) Line profiles of current for AgNP identified in the image in (**e**). (**g**) AFM topography image of hybrid Ag NP and ZnO NR on a metallic disc substrate. Scan sizes: (0.75 µm × 0.75 µm). (**h**) Current mapping of hybrid Ag NP and ZnO-NR. (**g**) Line profiles of current for AgNP and ZnO-NR identified in the image in (**i**).

In analyzing the reasons for the increase in the current at around 2.5 V, it is necessary to consider the intrinsic characteristics of ZnO nanorods and silver nanoparticles. ZnO is a wide-bandgap semiconductor, and in forward bias, significant current flows once the voltage exceeds the built-in potential barrier. The presence of silver nanoparticles, which are exceptional conductors, also aids in charge transport. The combination of ZnONRs with Ag-NPs can create a Schottky junction, improving carrier injection at the interface, and can slightly alter the turn-on voltage compared with individual components. The presence of Ag-NPs can introduce new surface states and modify existing defects in ZnONRs, impacting the overall conduction mechanism. Additionally, it can enhance the barrier height in reverse bias due to the Schottky effect, indicating improved stability and robustness of the device.

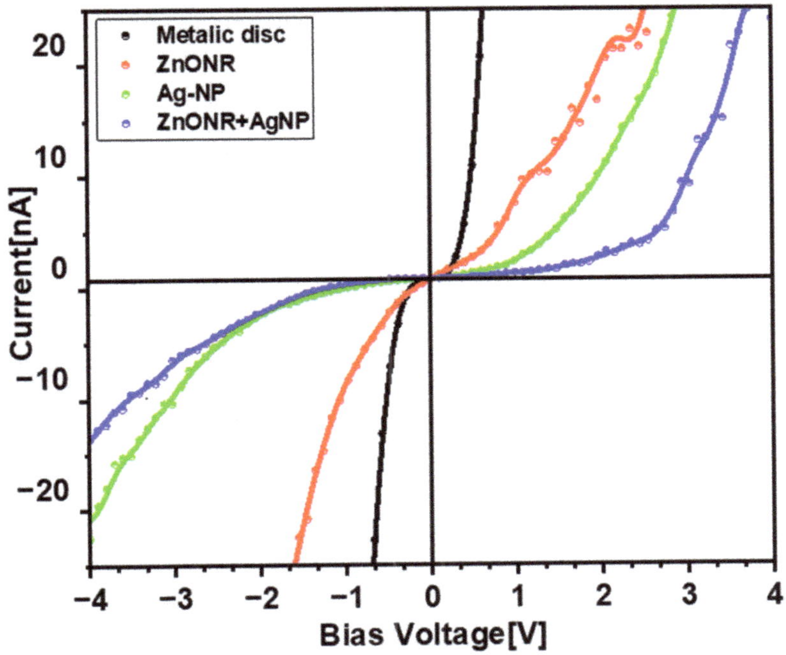

Figure 7. Local I-V characteristic of metallic disc (reference), ZnO nanorods, Ag nanoparticles, and hybrid ZnO-NRs and Ag NPs.

Furthermore, the presence of Ag-NPs can enhance the mobility of charge carriers in ZnONRs by providing additional pathways and reducing scattering [37]. It can also passivate surface defects, reducing trap states and enhancing charge transport. The interaction between Ag-NPs and ZnONRs can modify the band structure, leading to changes in electrical properties. It is important to mention that Schottky diodes can be made with higher threshold voltages using various semiconductor materials like silicon carbide (SiC) or gallium nitride (GaN) [38]. Additionally, Schottky diodes made with Au NPs and n-Si have a turn-on voltage of around 1.0 volt [39], indicating a noticeable difference in performance. On the other hand, the turn-on voltage for the Bi2S3 nanorods and Au nanoparticles device is approximately 0.72 volts [40], showing significant differences in voltage characteristics between different metal systems. However, the combination of ZnO nanorods and silver nanoparticles offers a more cost-effective synthesis option in this particular study.

3.4. Optical Properties of Hybrid of ZnO-NRs and Ag-NPs

Figure 8 illustrates the UV-Vis absorption spectra of ZnO nanorods (ZnO-NRs), silver nanoparticles (Ag-NPs), and their composites (ZnO-NR + Ag), providing important insights into their optical characteristics. The distinct absorption peak at 371 nm observed in the ZnO nanorods was in line with the near-band-edge absorption of ZnO-NR. Meanwhile, the silver nanoparticles showed a slightly shifted absorption peak at 371.5 nm, attributed to the localized surface plasmon resonance (LSPR) effect. The normal LSPR peak for silver nanoparticles is typically around 390–450 nm, so this shift to around 370 nm suggested that multiple factors such as nanoparticle size, dielectric environment, and surface chemistry may have been influencing the resonance behavior. Smaller nanoparticles tend to experience a blue shift in their plasmon resonance peaks due to quantum confinement effects, leading to absorption moving toward shorter wavelengths. The refractive index of the surrounding medium also plays a significant role, with a lower index resulting in a similar blue shift in the absorption peak. Additionally, surface modifications like ligand attachment or oxidation can impact the electronic properties of the nanoparticles, causing shifts in the LSPR peak [10,41–43]. The absorption peak for the composite was at 369.5 nm, slightly blue-shifted compared with the individual ZnO-NR and Ag-NP spectra. The analysis of energy gaps in the inset in Figure 8 shows the ZnO nanorods (ZnO-NRs), silver nanoparticles (AgNPs), and the ZnO-NR + AgNP hybrid, as illustrated in a Tauc plot, which plots the square root of the absorption coefficient against the photon energy. ZnO nanorods are known for their distinct composition and structure, displaying a noticeable absorption peak at the bandgap energy level, typically around 3.11 eV. In contrast, Ag nanoparticles exhibit absorption traits influenced by surface plasmon resonance, with a well-defined absorption peak near 3.18 eV [44].

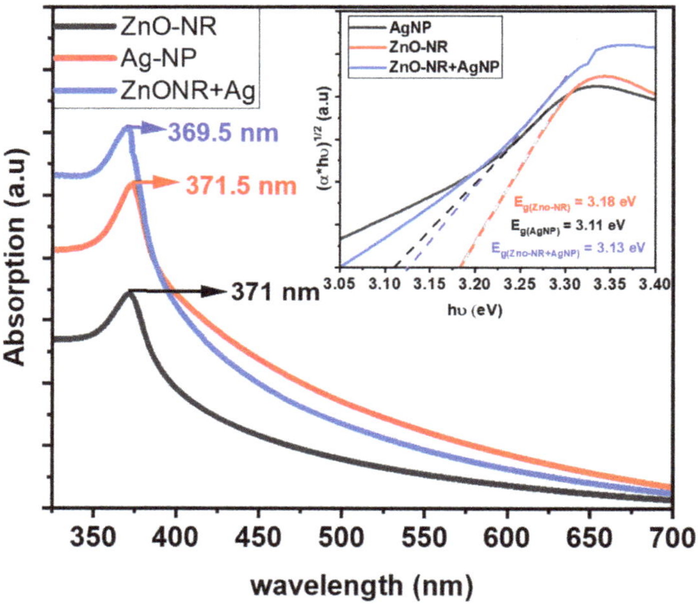

Figure 8. The absorption spectrum of ZnO nanorods, Ag nanoparticles, and hybrid ZnO-NR and Ag NPs. Inset figure: $(\alpha h\upsilon)^{1/2}$ vs. photon energy of ZnO nanorods, Ag nanoparticles, and hybrid ZnO-NRs and Ag NPs.

The alteration in the absorption spectrum when ZnO-NRs and AgNPs combined to form the ZnO-NR + AgNP hybrid was of particular interest. This change signified a close interaction between the ZnO nanorods and the Ag nanoparticles, resulting in a synergistic effect that enhanced the overall absorption performance. Specifically, the bandgap energy of the hybrid material experienced a slight shift, settling at approximately 3.13 eV. While seemingly small, this shift held great significance as it indicated the enhanced absorption capabilities of the hybrid material. The slight change in bandgap energy, approximately 3.13 eV, between the hybrid ZnO nanorods and the Ag nanoparticles was significant for several reasons. The surface plasmon resonance effect of Ag nanoparticles could enhance the electromagnetic field nearby and increase light absorption, impacting the properties of the ZnO nanorod. Additionally, the interaction between the two materials could result in charge transfer, altering the electronic structure of ZnO and causing a minor shift in bandgap energy. The size and distribution of Ag nanoparticles could also influence the electronic properties of ZnO, with quantum size effects coming into play due to the confinement of charge carriers at the nanoscale [45,46].

The photoluminescence spectra in Figure 9 show the presence of ZnO nanorods and Ag nanoparticles, with peaks at 384 nm and 386 nm, respectively. The addition of Ag nanoparticles at 379 nm had a significant effect on the optical characteristics of ZnO nanorods, resulting in increased photoluminescence intensity and changes in peak positions. This change was likely due to the plasmonic effects of the Ag nanoparticles and their interaction with the ZnO nanorods.

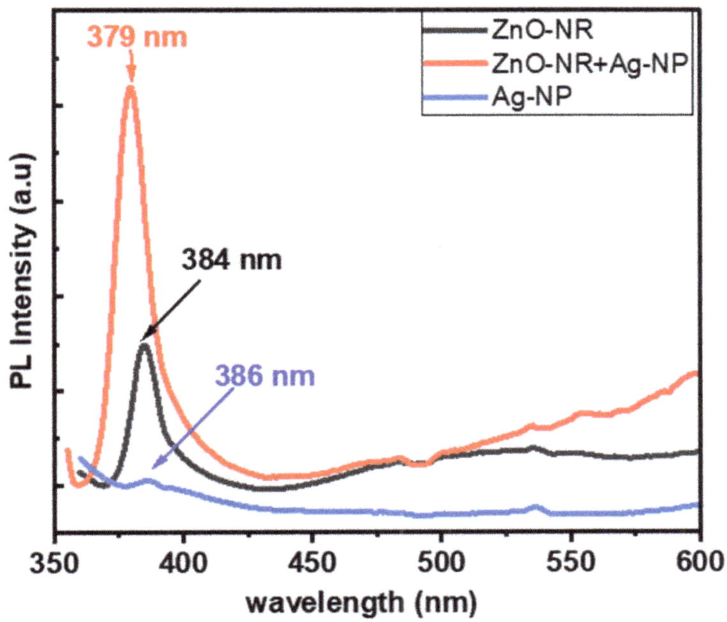

Figure 9. The photoluminescence spectrum o of ZnO nanorods, Ag nanoparticles, and hybrid ZnO-NRs and Ag NPs.

The addition of Ag nanoparticles caused a noticeable shift and change in intensity in the PL spectrum, with a shift from 379 nm to 386 nm, suggesting fundamental alterations in the electronic structure and energy states of ZnO nanorods. The heightened PL intensity for the hybrid structures indicates improved photoluminescence as a result of the strong interaction between ZnO nanorods and Ag nanoparticles.

One potential mechanism for the observed changes in PL intensity and peak positions is the surface plasmon resonance (SPR) of Ag nanoparticles, which can significantly enhance the local electromagnetic field and the PL intensity of ZnO nanorods. Another potential mechanism is the transfer of charge between the ZnO nanorods and Ag nanoparticles, which can affect the recombination rate of electron–hole pairs in ZnO, leading to variations in PL intensity and peak positions [47,48].

4. Conclusions

The detailed analysis of ZnO nanorods and silver nanoparticles provided valuable insights into their physical, structural, electrical, and optical characteristics. Utilizing scanning probe microscopy (SPM) and X-ray diffraction (XRD), it was determined that the ZnO nanorods exhibited a distinct shape, with lengths of around 350 nm and a highly crystalline, single-phase wurtzite structure. The silver nanoparticles, on the other hand, displayed a spherical shape with heights ranging from 5 to 14 nm and were confirmed to be composed of silver metal with a face-centered cubic structure. Kelvin probe force microscopy (KPFM) measurements offered further insights into the work function and contact potential differences (CPDs) of these nanomaterials. It was revealed that the work function of ZnO nanorods varied with size, ranging from approximately 4.556 eV to 4.527 eV, while the work function of Ag nanoparticles ranged from 4.56 eV to 4.538 eV. Conductive atomic force microscopy (C-AFM) was employed to investigate the electrical properties of these materials, uncovering regions of both high and low current in both the ZnO nanorods and the Ag nanoparticles. The hybrid ZnO-NR and Ag-NP material exhibited a unique electrical response, with a combined turn-on voltage of around 2.5 V. The optical properties of the ZnO-NR and Ag-NP hybrid were examined through absorption and photoluminescence (PL) spectra. The hybrid material demonstrated enhanced absorption capabilities, a shift in bandgap energy, and alterations in the photoluminescence peak and intensity. These findings highlight the positive impact of the hybridization of ZnO-NR and Ag-NP on their optical and electrical properties, positioning them as promising candidates for advanced nanotechnology applications.

Funding: This research received no external funding.

Data Availability Statement: The data that support the findings of this study are available from the corresponding author upon request.

Acknowledgments: The author thanks Palestine Technical University—Kadoorie (PTUK) and the Palestinian Ministry of Higher Education for facilities and support.

Conflicts of Interest: The author declare no conflict of interest.

References

1. Que, M.; Lin, C.; Sun, J.; Chen, L.; Sun, X.; Sun, Y. Progress in ZnO Nanosensors. *Sensors* **2021**, *21*, 5502. [CrossRef] [PubMed]
2. Wu, Y.; Ma, Y.; Zheng, H.; Ramakrishna, S. Piezoelectric Materials for Flexible and Wearable Electronics: A Review. *Mater. Des.* **2021**, *211*, 110164. [CrossRef]
3. Putri, A.E.; Roza, L.; Budi, S.; Ali Umar, A.; Fauzia, V. Tuning the Photocatalytic Activity of Nanocomposite ZnO Nanorods by Shape-Controlling the Bimetallic AuAg Nanoparticles. *Appl. Surf. Sci.* **2021**, *536*, 147847. [CrossRef]
4. Lefatshe, K.; Dube, P.; Sebuso, D.; Madhuku, M.; Muiva, C. Optical Dispersion Analysis of Template Assisted 1D-ZnO Nanorods for Optoelectronic Applications. *Ceram. Int.* **2021**, *47*, 7407–7415. [CrossRef]
5. Onyszko, M.; Zywicka, A.; Wenelska, K.; Mijowska, E. Revealing the Influence of the Shape, Size, and Aspect Ratio of ZnO Nanoparticles on Antibacterial and Mechanical Performance of Cellulose Fibers Based Paper. *Part. Part. Syst. Charact.* **2022**, *39*, 2200014. [CrossRef]
6. Boulahlib, S.; Dib, K.; Özacar, M.; Bessekhouad, Y. Optical, Dielectric, and Transport Properties of Ag-Doped ZnO Prepared by Aloe Vera Assisted Method. *Opt. Mater.* **2021**, *113*, 110889. [CrossRef]
7. Abutalib, M.M.; Rajeh, A. Influence of ZnO/Ag Nanoparticles Doping on the Structural, Thermal, Optical and Electrical Properties of PAM/PEO Composite. *Phys. B Condens. Matter* **2020**, *578*, 411796. [CrossRef]
8. Sharma, V.; Verma, D.; Okram, G.S. Influence of Surfactant, Particle Size and Dispersion Medium on Surface Plasmon Resonance of Silver Nanoparticles. *J. Phys. Condens. Matter* **2020**, *32*, 145302. [CrossRef]
9. Mehata, M.S. Green Route Synthesis of Silver Nanoparticles Using Plants/Ginger Extracts with Enhanced Surface Plasmon Resonance and Degradation of Textile Dye. *Mater. Sci. Eng. B* **2021**, *273*, 115418. [CrossRef]
10. Alzoubi, F.Y.; Ahmad, A.A.; Aljarrah, I.A.; Migdadi, A.B.; Al-Bataineh, Q.M. Localize Surface Plasmon Resonance of Silver Nanoparticles Using Mie Theory. *J. Mater. Sci. Mater. Electron.* **2023**, *34*, 2128. [CrossRef]
11. Juma, M.W.; Birech, Z.; Mwenze, N.M.; Ondieki, A.M.; Maaza, M.; Mokhotjwa, S.D. Localized Surface Plasmon Resonance Sensing of Trenbolone Acetate Dopant using Silver Nanoparticles. *Sci. Rep.* **2024**, *14*, 5721. [CrossRef]
12. Checa, M.; Neumayer, S.M.; Tsai, W.-Y.; Collins, L. Advanced Modes of Electrostatic and Kelvin Probe Force Microscopy for Energy Applications. In *Atomic Force Microscopy for Energy Research*; CRC Press: Boca Raton, FL, USA, 2022; ISBN 978-1-00-317404-2.
13. Wang, J.; Zhang, H.; Cao, G.; Xie, L.; Huang, W. Injection and Retention Characterization of Trapped Charges in Electret Films by Electrostatic Force Microscopy and Kelvin Probe Force Microscopy. *Phys. Status Solidi A* **2020**, *217*, 2000190. [CrossRef]
14. Li, X.; Hu, X.; Liu, M.; Sun, L.; Qiu, X. Electrical Characteristics of a Carbon Nanotube-Functionalized Probe for Kelvin Probe Force Microscopy. *J. Phys. Chem. C* **2020**, *124*, 28261–28266. [CrossRef]
15. Jakob, D.S.; Li, N.; Zhou, H.; Xu, X.G. Integrated Tapping Mode Kelvin Probe Force Microscopy with Photoinduced Force Microscopy for Correlative Chemical and Surface Potential Mapping. *Small* **2021**, *17*, 2102495. [CrossRef]
16. Hackl, T.; Poik, M.; Schitter, G. Heterodyne AC Kelvin Probe Force Microscopy for Nanoscale Surface Potential Imaging in Liquids. *IEEE Trans. Instrum. Meas.* **2023**, *72*, 1–8. [CrossRef]
17. Grévin, B.; Husainy, F.; Aldakov, D.; Aumaître, C. Dual-Heterodyne Kelvin Probe Force Microscopy. *Beilstein J. Nanotechnol.* **2023**, *14*, 1068–1084. [CrossRef]

18. Maryon, O.O.; Efaw, C.M.; DelRio, F.W.; Graugnard, E.; Hurley, M.F.; Davis, P.H. Co-Localizing Kelvin Probe Force Microscopy with Other Microscopies and Spectroscopies: Selected Applications in Corrosion Characterization of Alloys. *J. Vis. Exp.* **2022**, *184*, 64102. [CrossRef]
19. Lee, H.; Lee, W.; Lee, J.H.; Yoon, D.S. Surface Potential Analysis of Nanoscale Biomaterials and Devices Using Kelvin Probe Force Microscopy. *J. Nanomater.* **2016**, *2016*, 4209130. [CrossRef]
20. Beinik, I.; Kratzer, M.; Wachauer, A.; Wang, L.; Lechner, R.T.; Teichert, C.; Motz, C.; Anwand, W.; Brauer, G.; Chen, X.Y.; et al. Electrical Properties of ZnO Nanorods Studied by Conductive Atomic Force Microscopy. *J. Appl. Phys.* **2011**, *110*, 052005. [CrossRef]
21. Fernando, P.S.; Mativetsky, J.M. Unambiguous Measurement of Local Hole Current in Organic Semiconductors Using Conductive Atomic Force Microscopy. *J. Phys. Chem. C* **2023**, *127*, 9903–9910. [CrossRef]
22. Caballero-Quintana, I.; Amargós-Reyes, O.; Maldonado, J.-L.; Nicasio-Collazo, J.; Romero-Borja, D.; Barreiro-Argüelles, D.; Molnár, G.; Bousseksou, A. Scanning Probe Microscopy Analysis of Nonfullerene Organic Solar Cells. *ACS Appl. Mater. Interfaces* **2020**, *12*, 29520–29527. [CrossRef]
23. Khoury, R.; Alvarez, J.; Ohashi, T.; Martín, I.; Ortega, P.; Lopez, G.; Jin, C.; Li, Z.; Rusli; Bulkin, P.; et al. Observation of Photovoltaic Effect within Locally Doped Silicon Nanojunctions Using Conductive Probe AFM. *Nano Energy* **2020**, *76*, 105072. [CrossRef]
24. Barrigón, E.; Hrachowina, L.; Borgström, M.T. Light Current-Voltage Measurements of Single, as-Grown, Nanowire Solar Cells Standing Vertically on a Substrate. *Nano Energy* **2020**, *78*, 105191. [CrossRef]
25. Musa, I.; Qamhieh, N.; Mahmoud, S.T. Synthesis and Length Dependent Photoluminescence Property of Zinc Oxide Nanorods. *Results Phys.* **2017**, *7*, 3552–3556. [CrossRef]
26. Musa, I.; Mousa, R. Synthesis and Characterization of Variable-Sized Silver Nanoparticles Using Pistacia Palaestina Leaf Extract. *Plasmonics* **2024**, 1–9. [CrossRef]
27. Meena, P.L.; Bhardwaj, P.; Kumar, Y.; Singh, S.P. Synthesis and Characterization of One Dimensional ZnO Nanorods. In Proceedings of the AIP Conference Proceedings, Longowal, India, 9–11 November 2021; Volume 2352, p. 040043. [CrossRef]
28. Musa, I.; Qamhieh, N. Study of optical energy gap and quantum confinment effects in zinc oxide nanoparticles and nanorods. *Dig. J. Nanomater. Biostruct.* **2019**, *14*, 119–125.
29. Ali, H.; Azad, A.K.; Khan, K.A.; Rahman, O.; Chakma, U.; Kumer, A. Analysis of Crystallographic Structures and Properties of Silver Nanoparticles Synthesized Using PKL Extract and Nanoscale Characterization Techniques. *ACS Omega* **2023**, *8*, 28133–28142. [CrossRef]
30. Glatzel, T.; Gysin, U.; Meyer, E. Kelvin Probe Force Microscopy for Material Characterization. *Microscopy* **2022**, *71*, i165–i173. [CrossRef] [PubMed]
31. Musa, I.; Faqi, R. Structural, Electrostatic Force Microscopy, Work Function, and Optical Characterization of Pure and Al-Doped ZnO Nanoparticles. *Results Mater.* **2024**, *22*, 100570. [CrossRef]
32. Mishra, Y.K.; Chakravadhanula, V.S.K.; Hrkac, V.; Jebril, S.; Agarwal, D.C.; Mohapatra, S.; Avasthi, D.K.; Kienle, L.; Adelung, R. Crystal Growth Behaviour in Au-ZnO Nanocomposite under Different Annealing Environments and Photoswitchability. *J. Appl. Phys.* **2012**, *112*, 064308. [CrossRef]
33. Sahana, M.B.; Sudakar, C.; Dixit, A.; Thakur, J.S.; Naik, R.; Naik, V.M. Quantum Confinement Effects and Band Gap Engineering of SnO_2 Nanocrystals in a MgO Matrix. *Acta Mater.* **2012**, *60*, 1072–1078. [CrossRef]
34. Musa, I.; Qamhieh, N.; Mahmoud, S.T. Ag Nanocluster Production through DC Magnetron Sputtering and Inert Gas Condensation: A Study of Structural, Kelvin Probe Force Microscopy, and Optical Properties. *Nanomaterials* **2023**, *13*, 2758. [CrossRef]
35. Makov, G.; Nitzan, A.; Brus, L.E. On the Ionization Potential of Small Metal and Dielectric Particles. *J. Chem. Phys.* **1988**, *88*, 5076–5085. [CrossRef]
36. Musa, I.; Ghabboun, J. Work Function, Electrostatic Force Microscopy, Tunable Photoluminescence of Gold Nanoparticles, and Plasmonic Interaction of Gold Nanoparticles/Rhodamine 6G Nanocomposite. *Plasmonics* **2024**, 1–10. [CrossRef]
37. Thappily, P.; Mandin, P.; Sauvage, T.; Sandhya, K. Enhancement of Charge Transport Properties of a Novel Rubbery Semiconductor via Silver Nanocomplexing. *Mater. Sci. Semicond. Process.* **2021**, *131*, 105854. [CrossRef]
38. Kumar, A.; Moradpour, M.; Losito, M.; Franke, W.-T.; Ramasamy, S.; Baccoli, R.; Gatto, G. Wide Band Gap Devices and Their Application in Power Electronics. *Energies* **2022**, *15*, 9172. [CrossRef]
39. Abbas, Y.; Rezk, A.; Anwer, S.; Saadat, I.; Nayfeh, A.; Rezeq, M. Improved Figures of Merit of Nano-Schottky Diode by Embedding and Characterizing Individual Gold Nanoparticles on n-Si Substrates. *Nanotechnology* **2020**, *31*, 125708. [CrossRef]
40. Saha, S.K.; Pal, A.J. Schottky Diodes between Bi_2S_3 Nanorods and Metal Nanoparticles in a Polymer Matrix as Hybrid Bulk-Heterojunction Solar Cells. *J. Appl. Phys.* **2015**, *118*, 014503. [CrossRef]
41. Link, S.; El-Sayed, M.A. Size and Temperature Dependence of the Plasmon Absorption of Colloidal Gold Nanoparticles. *J. Phys. Chem. B* **1999**, *103*, 4212–4217. [CrossRef]
42. Wu, Q.; Si, M.; Zhang, B.; Zhang, K.; Li, H.; Mi, L.; Jiang, Y.; Rong, Y.; Chen, J.; Fang, Y. Strong Damping of the Localized Surface Plasmon Resonance of Ag Nanoparticles by Ag_2O. *Nanotechnology* **2018**, *29*, 295702. [CrossRef]
43. Mulvaney, P. Surface Plasmon Spectroscopy of Nanosized Metal Particles. *Langmuir* **1996**, *12*, 788–800. [CrossRef]
44. Campos, A.; Troc, N.; Cottancin, E.; Pellarin, M.; Weissker, H.-C.; Lermé, J.; Kociak, M.; Hillenkamp, M. Plasmonic Quantum Size Effects in Silver Nanoparticles Are Dominated by Interfaces and Local Environments. *Nat. Phys.* **2019**, *15*, 275–280. [CrossRef]

45. Mandal, S.; Ananthakrishnan, R. Double Effects of Interfacial Ag Nanoparticles in a ZnO Multipod@Ag@Bi_2S_3 Z-Scheme Photocatalytic Redox System: Concurrent Tuning and Improving Charge-Transfer Efficiency. *Inorg. Chem.* **2020**, *59*, 7681–7699 [CrossRef] [PubMed]
46. Liu, C.; Xu, X.; Wang, C.; Qiu, G.; Ye, W.; Li, Y.; Wang, D. ZnO/Ag Nanorods as a Prominent SERS Substrate Contributed by Synergistic Charge Transfer Effect for Simultaneous Detection of Oral Antidiabetic Drugs Pioglitazone and Phenformin. *Sens. Actuators B Chem.* **2020**, *307*, 127634. [CrossRef]
47. Wang, X.; Ye, Q.; Bai, L.-H.; Su, X.; Wang, T.-T.; Peng, T.-W.; Zhai, X.-Q.; Huo, Y.; Wu, H.; Liu, C.; et al. Enhanced UV Emission from ZnO on Silver Nanoparticle Arrays by the Surface Plasmon Resonance Effect. *Nanoscale Res. Lett.* **2021**, *16*, 8. [CrossRef]
48. Rauwel, P.; Galeckas, A.; Rauwel, E. Enhancing the UV Emission in ZnO–CNT Hybrid Nanostructures via the Surface Plasmon Resonance of Ag Nanoparticles. *Nanomaterials* **2021**, *11*, 452. [CrossRef]

surfaces

MDPI

Article

Evaluation of Photocatalytic Hydrogen Evolution in Zr-Doped TiO_2 Thin Films

Luis F. Garay-Rodríguez [1,*], M. R. Alfaro Cruz [1,2], Julio González-Ibarra [1], Leticia M. Torres-Martínez [1,3,*] and Jin Hyeok Kim [4]

1 Departamento de Ecomateriales y Energía, Facultad de Ingeniería Civil, Universidad Autónoma de Nuevo León, Cd. Universitaria, San Nicolás de los Garza 66455, Mexico; malfaroc@uanl.edu.mx (M.R.A.C.); julio.gonzalezibr@uanl.edu.mx (J.G.-I.)
2 Departamento de Ecomateriales y Energía, Facultad de Ingeniería Civil, CONAHCYT-Universidad Autónoma de Nuevo León, Cd. Universitaria, San Nicolás de los Garza 66455, Mexico
3 Centro de Investigación en Materiales Avanzados, S.C., Miguel de Cervantes 120 Complejo Industrial Chihuahua, Chihuahua 31110, Mexico
4 Department of Materials Science and Engineering and Optoelectronics Convergence Research Center, Chonnam National University, Yongbong-ro 77, Buk-gu, Gwangju 61186, Republic of Korea; jinhyeok@chonnam.ac.kr
* Correspondence: lgarayr@uanl.edu.mx or lfgarayr@gmail.com (L.F.G.-R.); lettorresg@yahoo.com or leticia.torres@uanl.edu.mx (L.M.T.-M.)

Abstract: Doping titanium dioxide has become a strategy for enhancing its properties and reducing its recombination issues, with the aim of increasing its efficiency in photocatalytic processes. In this context, this work studied its deposition over glass substrates using a sol–gel dip coating methodology. The effect of doping TiO_2 with Zirconium cations in low molar concentrations (0.01, 0.05, 0.1%) in terms of its structural and optical properties was evaluated. The structural characterization confirmed the formation of amorphous thin films with Zr introduced into the TiO_2 cell (confirmed by XPS characterization), in addition to increasing and defining the formed particles and their size slightly. These changes resulted in a decrease in the transmittance percentage and their energy band gap. Otherwise, their photocatalytic properties were evaluated in hydrogen production using ethanol as a sacrificial agent and UV irradiation. The hydrogen evolution increased as a function of the Zr doping, the sample with the largest Zr concentration (0.1% mol) being the most efficient, evolving 38.6 mmolcm^{-2} of this gas. Zr doping favored the formation of defects in TiO_2, being responsible for this enhancement in photoactivity.

Keywords: TiO_2; thin films; H_2 production; photocatalysis

1. Introduction

It is known that titanium dioxide as a photocatalyst has been widely used in different environmental applications mainly due to its chemical stability, low cost, and its adequate band position to perform simultaneous red-ox reactions [1]. On the other hand, some of its optical properties make it an ideal candidate to be used as a thin film [2]. In this context, using it as a thin film can maximize light absorption because of the maximization of the surface area used in the material [3]. Unfortunately, one of its most common issues is the fast recombination of photo-generated charges after its illumination. For this purpose, many strategies have been carried out to avoid or minimize this phenomenon, improving its physicochemical properties to enhance its reaction yields.

Metal doping has resulted as an efficient alternative to achieve these goals; moreover, in most cases, its band gap energy can be reduced so it can be activated in the visible-light region. Many reports have shown the effect of metal doping in photocatalytic processes, highlighting the use of some elements such as Cu- [4,5], Ag- [6,7], Au- [8,9], Pt- [10,11] Ni- [12,13], among others.

In this context, Zr is a good candidate for doping TiO_2 lattices because both elements are in the same group, have the same valence state, and the anatase phase supports Zr incorporation, forming the solid solution $Ti_{1-x}Zr_xO_2$ [14]. Some reports have studied the effect of Zr introduction into the TiO_2 anatase lattice on its physicochemical properties. For instance, Bolbol et al. [14] deposited Zr-doped TiO_2 thin films over glass substrates, varying the Zr dopant concentration from 0.5 to 10% mol using the sol–gel spin coating technique. They found changes in the structural and optical properties owing to a micro-strain increase created by Zr addition. This feature caused a reduction in electron–hole recombination compared to pristine TiO_2. Similarly, Juma et al. [15] and Oluwabi et al. [16] explored Zr-TiO_2 samples deposited by spray pyrolysis. In this context, Zr introduction in the TiO_2 lattice suppressed the anatase in the rutile phase transformation process, reduced the film roughness, decreased the film's crystallinity, and strongly increased the dielectric constant. In photocatalytic applications, Mbiri et al. [17] evaluated the Zr dopant content effect on TiO_2 thin films in the degradation of persistent organic pollutants, such as Chlorisazon, Phenol, and 4-Chlorophenol, finding a reduction in the recombination rate and achieving degradation percentages higher than 80%. Similar findings were reported by degrading formaldehyde [18], methylene blue [19–21], 4-Nitrophenol [22], bismark brown red [23], methylene orange [24], and 4-chlorophenol [25]. On the other hand, fewer reports have been observed in the case of hydrogen evolution; for instance, Chattopadhyay et al. [26]. prepared $Ti_{1-x}Zr_xO_{2-y}$ nanocrystals in different Zr compositions. They found that Zr^{4+} incorporation into the TiO_2 lattice modified the surface chemistry, caused lattice strain and increased the amount of Ti^{3+} species that favored the electron transference, reaching a superior hydrogen production compared with that of pristine TiO_2. Some other works have reported the effect of Zr dopant in TiO_2; however, most of them focus on the use of powder particles [27–30].

Considering the above, this work reports the findings of the effect of Zr doping on TiO_2 thin films deposited over a glass substrate by sol–gel dip coating in terms of their optical and structural properties. Also, the impact of these changes on the photocatalytic efficiency in the hydrogen evolution reaction using ethanol as sacrificial agent is described.

2. Materials and Methods

2.1. Thin Film Deposition

Pristine and Zr-doped TiO_2 thin films were deposited by the sol–gel chemical method. In this context, a Titanium butoxide (97% Sigma Aldrich, St. Louis, MO, USA) 3 M solution was prepared in isopropanol (DEQ), adding the appropriate %mol of Zirconium butoxide (97% Sigma Aldrich) to obtain 0.01, 0.05, or 0.1 mol of Zr in the media. All solutions were deposited over glass substrates (previously washed in separate isopropanol-acetone-water washings) using a dip coating technique. For this purpose, the glass substrates were placed vertically in the system and immersed three times into the solution at a constant speed. After each immersion cycle, the solvent was evaporated instantly by subjecting the substrates to a hot temperature (165 °C). Additionally, the films were calcined at 400 °C for 2 h to promote phase crystallization.

2.2. Characterization

The structural characterization of the films was evaluated using an X-ray PANanalytical diffractometer with Cu Ka 1.54 Å radiation at a grazing incidence angle. Surface images were taken in ASYLUM RESEARCH MFP3D-SA AFM equipment in tapping mode. Transmittance spectra were obtained with a UV–VIS/Cary 5000, running the samples at a 200–800 nm wavelength interval, and PL spectra were measured in an Agilent Cary Eclipse (excitation wavelength—325 nm) fluorescence spectrophotometer. XPS measurements were analyzed using a VG Multilab 2000 (Thermo VG Scientific equipment, Waltham, MA, USA) with a monochromatic MG-Kα (1253.6 eV) irradiation source.

2.3. Photocatalytic Reactions

Photocatalytic hydrogen production reactions were performed under UV (254 nm) irradiation using a cylindrical Pyrex batch reactor. Four films (30 cm^2 of active area) were pasted inside the reactor, adding 200 mL of deionized water. The system was vented with Argon for 15 min to promote an anoxic media, and then a UV lamp was immersed through a quartz tube and turned on. Gas samples were taken every 30 min using a syringe and injected in a gas chromatograph Varian GP-3380 with a thermal conductivity detector using Argon as the mobile phase and RESTEK REST-19808 (RESTEK, Centre County, PA, USA) column as a stationary phase. The acetaldehyde concentration in the remaining reaction liquid was measured by High-Resolution Liquid Chromatography using a Shimadzu Nexcol C18 (Shimadzu, Columbia, MD, USA) column as the stationary phase, and a mixture of acetonitrile/water 45:55 solution as the mobile phase.

3. Results and Discussion

Figure 1 presents the XRD patterns of all the samples. As seen, the pristine TiO_2 and the Zr-doped films were amorphous, and no diffraction peaks were detected. Similar results have been reported in the literature for TiO_2 films [31–34], where peaks with very low intensities or no peaks are detected in similar sol–gel deposition conditions.

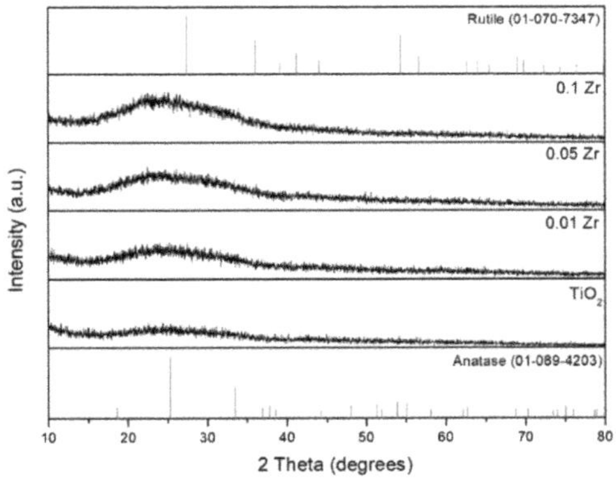

Figure 1. XRD patterns of the deposited Zr-doped TiO_2 thin films.

Figure 2 shows the elemental composition of the TiO_2 and Zr-doped TiO_2 films according to the XPS spectra for the C 1s, O 1s, Ti 2p, and Zr 3d levels. The C 1s core level has been deconvoluted into three peaks related to C-C, C-O-C, and O-C=O bonds (Figure 2a). On the other hand, the O1s spectra were deconvoluted into three curves for the TiO_2 and 0.01 Zr films and four curves for the 0.05 Zr and 0.1 Zr films. These curves correspond to the M-O, M-OH, C-OH, and C-O bonds [35,36], with only the 0.05 Zr and 0.1 Zr films showing the contribution of the C-OH species [35]. Due to the Zr being incorporated, the peak related to the M-O bond increased; likewise, the 0.01 Zr film is the only one that has less area in the peaks related to the M-OH and C-O bonds (Figure 2b).

Additionally, the Ti2p region is contributed to by the Ti^{3+}, Ti^{4+}, and $Ti(OH)_2$ (marked with *) species [36–38], while the Zr3d core level shows the presence of Zr^{4+} species [17,36]. In this context, the binding energy values obtained in the Zr 3d spectra are slightly lower to the ones reported in the ZrO_2 spectra, which suggests that the atoms are incorporated into the TiO_2 structure instead of in the ZrO_2 phase in low proportions [15]. Additionally, as the electronegativity of the Ti (1.54) is higher than that of the Zr (1.33), the peaks in the Ti2p and Zr3d spectra shifted toward a lower binding energy between the samples with Zr^{4+}, which is related to the partial substitution of Ti^{4+} by Zr^{4+} ions [15,36]. This feature can be related to the formation of a $Ti_{1-x}Zr_xO_2$ phase.

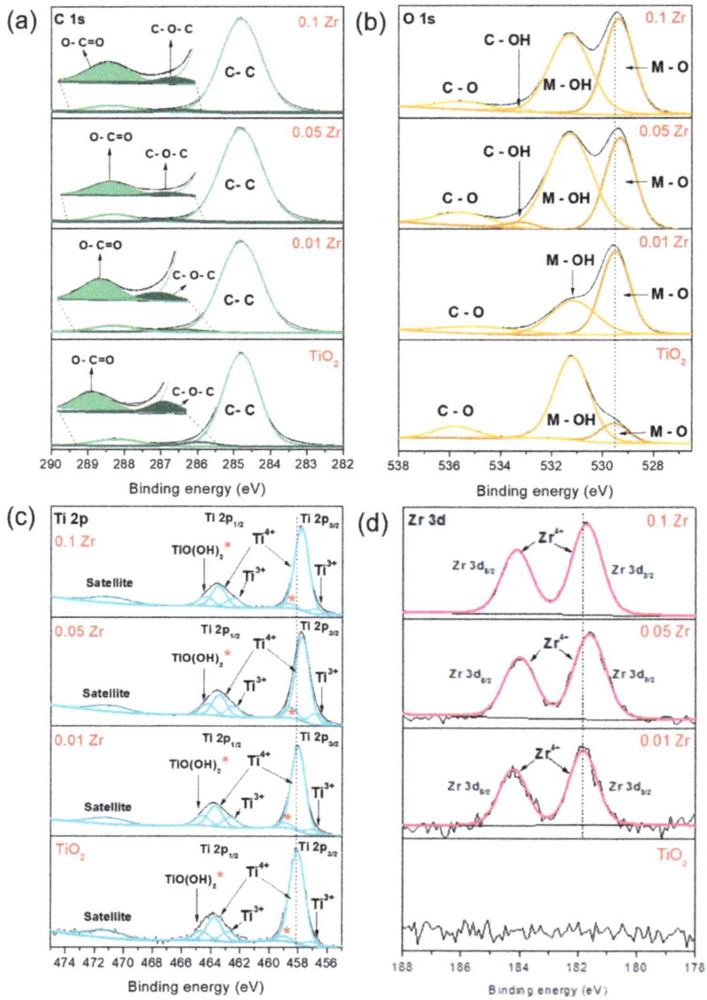

Figure 2. XPS core levels of (**a**) C 1s, (**b**) O 1s, (**c**) Ti 2p, and (**d**) Zr 3d.

Figure 3 presents the AFM images taken in a contact mode. As seen, all films were deposited uniformly; however, the pristine TiO_2 sample presented the formation of fine particles with the appearance of some cracks potentially generated during solvent evaporation. In contrast, the incorporation of Zr promoted the formation of densely packed particles, and additionally it reduced the formation of cracks, promoting a good coverage of the substrate. According to some authors, Zr incorporation as a dopant can retard the TiO_2 densification, reducing the formation of cracks and pinholes in the layer [39,40]. On the other hand, similarity can be seen between the TiO_2, 0.01 Zr, and 0.05 Zr samples; however, a slight reduction in the surface roughness was observed (Figure 3). Similar findings were reported by Naumenko et al. [39], where low Zr loads promoted the formation of densely packed particles as a result of the reduced crystallization effect that Zr promotes, or the increase in the nucleation centers during the film growth which inhibits grain growth, as Juma et al. mentions [15].

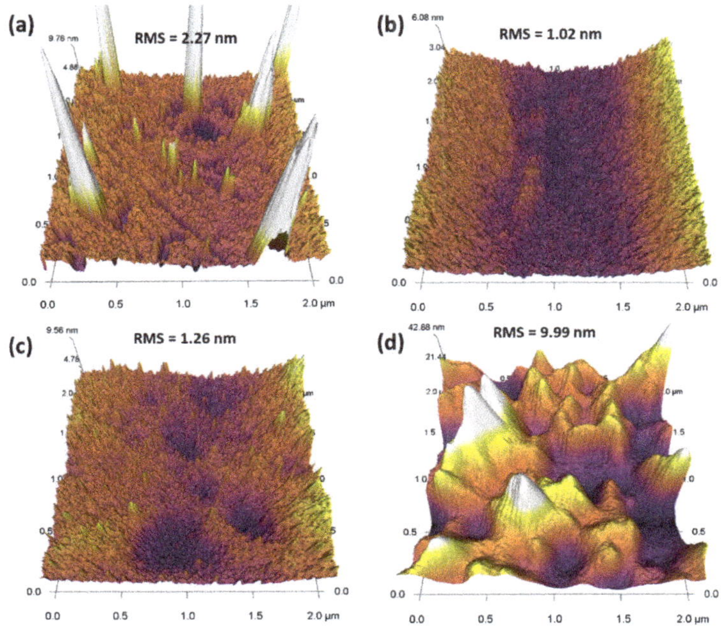

Figure 3. AFM characterization of the deposited films. (**a**) TiO_2; (**b**) 0.01 Zr; (**c**) 0.05 Zr; (**d**) 0.1 Zr.

On the contrary, a larger Zr concentration (0.1 Zr sample) promotes an increase in the film surface roughness; in this context, this drastic change can be associated with the formation of a more viscous sol due to the addition of a larger Zr precursor concentration [17]. An increase in the surface roughness can suggest an increase in the surface area, resulting in the presence of more active sites to enhance their photoactivity [41].

The transmittance percentage obtained from the deposited films is presented in Figure 4a. All the films are transparent in the visible light region (% T > 70%); however, when the Zr^{4+} concentration increases, the transmittance percentage slightly decreases, this behavior being related to high light dispersion over the film surface due to the increased roughness (more remarkable in the 0.1 Zr sample) [42,43]. The optical band gap of the films was calculated by the Tauc plot using the following equation [44]:

$$(\alpha h\nu) = A\left(h\nu - E_g\right)^n$$

where α is the absorption coefficient, $h\nu$ is the photon energy, A is a proportionality constant, n is the Tauc exponent ($n = 1/2$ for direct transitions and $n = 2$ for indirect transitions), and E_g is the band gap of the material [44,45]. Figure 4b shows a Tauc plot, where the band gap values slightly decrease with the incorporation of Zr (TiO_2 = 3.88 eV; 0.01 Zr = 3.81 eV; 0.05 Zr = 3.85 eV; 0.1 Zr = 3.76 eV). This decrease in the band gap value is related to shifts in the absorption edge toward a higher wavelength, which is due to an increase in the doping carrier concentrations. These carriers interact with free carriers and ionized impurities, causing a decrease in the band gap value [46]. In our films, as the Zr^{4+} concentration increases, the absorption edge shifts to higher wavelengths, decreasing the band gap value due to the increase in different impurities from the Zr^{4+}. According to Bolbol et al., this phenomenon is known as the Burstein–Moss effect [14].

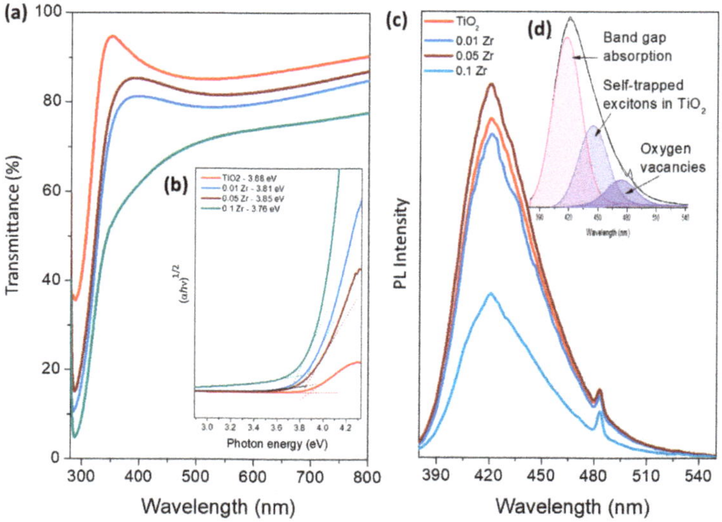

Figure 4. Optical properties as (**a**) transmittance percentage, (**b**) band gap (Tauc plot), (**c**) PL spectra, and (**d**) PL spectrum deconvoluted from the TiO_2 and Zr-doped TiO_2 films.

The PL spectra of the films excited with a wavelength of 320 nm are shown in Figure 4c. As seen, all the samples presented a broad emission between 390 and 540 nm. As this band displays a broad emission, this region could be deconvoluted into three bands, which are related to the TiO_2 band gap absorption (∼420 nm), self-trapped excitons in TiO_2 (∼450 nm), and oxygen vacancies of TiO_2 and of ZrO_2 and Zr^{4+} (∼430 to 530 nm) [47–49] (Figure 4d). On the other hand, from Figure 4c, there is an evident decrease in the PL intensity in the samples with a larger load of Zr^{4+}; in this context, it is well known that the PL emission reduction is associated with a minimization of the e^-/h^+ recombination, which translates into a greater availability of free charges to carry out red-ox reactions [50]. Additionally, it is known that the doping process produces extra free electrons in the TiO_2 lattice, which also reduces the emission efficiency by creating a non-radiative channel [14].

The photocatalytic hydrogen evolution rates obtained from the deposited films using ethanol as a sacrificial agent are presented in Figure 5a. As seen, all the samples present photoactivity, which is enhanced because of the increase in the Zr^{4+} concentration, reaching almost double the TiO_2 production (22 mmolcm^{-2}) in the sample loaded with 0.1%mol of Zr (38.6 mmolcm^{-2}). In this context, this observed improvement in the photocatalytic activity can be related to different facts; for instance, a larger Zr concentration causes a greater roughness on the film surface, which, as mentioned previously, increases the surface area and the active sites where ethanol molecules can be adsorbed and react to produce hydrogen. Additionally, the 0.1 Zr sample presented reduced electron–hole recombination, evidenced by the PL analyses, mainly associated with the formation of energy levels below the TiO_2 conduction band as self-trapped excitons, and defects such as oxygen vacancies.

Figure 5b presents the possible mechanism of hydrogen production through ethanol oxidation. In this context, the use of this organic compound has been highlighted by its capacity to act as an electron donor to the conduction band, enhancing the hydrogen production compared with using pure water [29]. Additionally, the presence of the different defects can capture light-induced electrons; for instance, in the case of oxygen vacancies, they play an important role acting as electron (e^-) traps, which, in consequence, avoid their recombination with the holes (h^+) [51]. Moreover, Zr^{4+} doping also introduces alternative defect levels close to the conduction band of TiO_2, which also can act as electron trap centers because the presence of Zr metal is used to produce a Schottky barrier to facilitate electron capture [52].

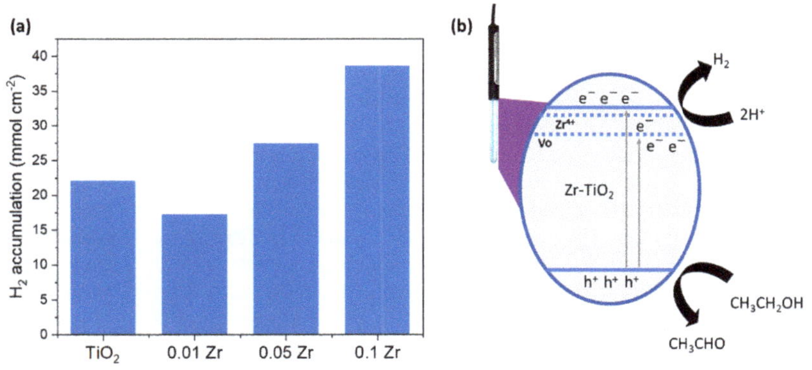

Figure 5. (**a**) Photocatalytic hydrogen production rate as function of the surface area. (**b**) Proposed mechanism of the hydrogen evolution using Zr-doped TiO_2.

On the other hand, ethanol is oxidated in the Zr-TiO_2 film valence band, forming acetaldehyde. The formation of this compound was evidenced in the remaining liquid of the reaction, detecting the following concentrations: TiO_2 = 750 μmol, 0.01 Zr = 627 μmol, 0.05 Zr = 612 μmol, 0.1 Zr = 530 μmol. Finally, the protons (H^+) formed from ethanol oxidation react with the electrons (e^-) accumulated in the formed defects and the conduction band to efficiently produce hydrogen.

Finally, Table 1 presents a summary of the hydrogen evolution reaction via photocatalysis or photo-electrocatalysis using TiO_2 thin films deposited under different methodologies. In this context, some studies focus on analyzing the different deposition conditions related to the used methodology, while others analyze the effect of an added dopant concentration (added in situ or as a multi-layer). As seen, the hydrogen production values obtained in this work are comparable with those that present higher production values, which suggests that doping TiO_2 with this metal (Zr) is an alternative to enhance its photocatalytic performance.

Table 1. Summary of the photocatalytic hydrogen production studies using TiO_2 thin films.

Photocatalyst	Dopant	Deposition Method	Illumination	Sacrificial Agent	Hydrogen Production	Ref.
TiO_2	Cu-Ni	Drop casting	300 W Xe lamp	Methanol 25%	41,690 μmol/gh	[53]
TiO_2	--	DC sputtering	UV 254 nm lamp	--	38 μmol	[54]
TiO_2	--	Hydrothermal	UV 254 nm lamp	--	132 μmol	[55]
TiO_2	--	RF sputtering	300 W Xe lamp	--	0.55 μmol/hcm^2	[56]
TiO_2	Au-Pd	Spin coating	300 W Xe lamp	Glycerol 5%	0.014 mL/min	[57]
TiO_2	Ag	Sol–gel/dip coating	5000 W Xe lamp	KOH 1 M	580 μmol	[58]
TiO_2	Ag	Drop casting	420 W Hg lamp	Water/methanol 1:1	148 μmol/gh	[59]
TiO_2	Cr	RF sputtering	250 W W lamp	NaOH 1 M	24 μmol/h	[60]
TiO_2	Ag	Hydrothermal	16 W Hg lamp	Ethanol 10%	8.1 μmol/cm^2	[61]
TiO_2	Pt	Dip coating	Black light lamps	Ethanol 50%	9 μmol/min	[62]
TiO_2	Pt	RF sputtering Sol–gel/spin coating	250 W W lamp	NaOH 1M	12.5 μmol/h 4.3 μmol/h	[63]
TiO_2	N-NiO N-CuO	Sol–gel/dip coating	UV 254 nm lamp	--	62,000 μmol/g	[64]
TiO_2	Pt	Dip coating	13 W UV lamp	Water/methanol 1:1	349.6 μmol/gh	[65]
TiO_2	Zr	Anodization method	500 W Xe lamp	Artificial sea water/ethilenglycol	15 μmol	[30]
TiO_2	Zr	Sol–gel/dip coating	UV 254 lamp	ethanol	38,600 μmol/cm^2	This work

4. Conclusions

Zr-doped TiO_2 amorphous thin films were successfully grown over glass substrates using the sol–gel dip coating methodology, varying the Zr concentration in the film precursor solution (0.01, 0.05 and 0.1 mol%). The incorporation of Zr^{4+} into the TiO_2 cell was confirmed using XPS, which resulted in the possible formation of the $Ti_{1-x}Zr_xO_2$ phase. The increase in the Zr concentration caused changes in the structural and optical properties of the films such as an increase in the film roughness, and a slight reduction in the transmittance percentage and their band gap. All the films exhibited photoactivity, evolving hydrogen under UV irradiation and ethanol as a sacrificial agent, with the hydrogen accumulation being increased by almost double in the samples with the largest Zr load (0.1 Zr = 38.6 mmolcm^{-2}) compared with that of pristine TiO_2 film. This enhancement in the photocatalytic activity was associated with the increase in the roughness of these films, which resulted in an increased surface area and was favorable for the reaction, and with the formation of different defects near to the TiO_2 conduction band that acted as electron trap centers and minimized the e^-/h^+ recombination rate.

Author Contributions: L.F.G.-R. and M.R.A.C.: conceptualization, methodology, investigation, data curation, writing—original draft. J.G.-I.: methodology, data curation. L.M.T.-M.: resources, supervision, writing—review and editing. J.H.K.: resources, supervision. All authors have read and agreed to the published version of the manuscript.

Funding: This research was founded by Consejo Nacional de Humanidades, Ciencias y Tecnologías: 320379; Universidad Autónoma de Nuevo León: PAICYT 275-CE-2022; PAICYT 277-CE-2022; National Research Foundation of Korea: 2022R1A2C2007219.

Institutional Review Board Statement: Not applicable.

Informed Consent Statement: Not applicable.

Data Availability Statement: The data that support the findings of this study are available from the corresponding author upon reasonable request.

Acknowledgments: The authors express their gratitude to the Centro de Investigación en Materiales Avanzados (CIMAV), specially Luz Ibarra and Oscar Vega for their valuable assistance with the AFM characterization, and to the Chonnam National University for the XPS measurement. Julio González thanks CONAHCYT for his 1148572 "Ayudante de Investigador" grant.

Conflicts of Interest: The authors declare no conflicts of interest.

References

1. Pant, B.; Park, M.; Park, S.-J. Recent Advances in TiO_2 Films Prepared by Sol-Gel Methods for Photocatalytic Degradation of Organic Pollutants and Antibacterial Activities. *Coatings* **2019**, *9*, 613. [CrossRef]
2. Alam, M.J.; Cameron, D.C. Preparation and Characterization of TiO_2 Thin Films by Sol-Gel Method. *J. Solgel Sci. Technol.* **2002**, *25*, 137–145. [CrossRef]
3. Nalajala, N.; Patra, K.K.; Bharad, P.A.; Gopinath, C.S. Why the Thin Film Form of a Photocatalyst Is Better than the Particulate Form for Direct Solar-to-Hydrogen Conversion: A Poor Man's Approach. *RSC Adv.* **2019**, *9*, 6094–6100. [CrossRef] [PubMed]
4. Liu, L.; Gao, F.; Zhao, H.; Li, Y. Tailoring Cu Valence and Oxygen Vacancy in Cu/TiO_2catalysts for Enhanced CO_2photoreduction Efficiency. *Appl. Catal. B* **2013**, *134–135*, 349–358. [CrossRef]
5. Colón, G.; Maicu, M.; Hidalgo, M.C.; Navío, J.A. Cu-Doped TiO_2 Systems with Improved Photocatalytic Activity. *Appl. Catal. B* **2006**, *67*, 41–51. [CrossRef]
6. Zhao, C.; Krall, A.; Zhao, H.; Zhang, Q.; Li, Y. Ultrasonic Spray Pyrolysis Synthesis of Ag/TiO_2 Nanocomposite Photocatalysts for Simultaneous H_2 Production and CO_2 Reduction. *Int. J. Hydrogen Energy* **2012**, *37*, 9967–9976. [CrossRef]
7. Li, H.; Wu, X.; Wang, J.; Gao, Y.; Li, L.; Shih, K. Enhanced Activity of AgMgOTiO_2 Catalyst for Photocatalytic Conversion of CO_2 and H_2O into CH_4. *Int. J. Hydrogen Energy* **2016**, *41*, 8479–8488. [CrossRef]
8. Puga, A.V.; Forneli, A.; García, H.; Corma, A. Production of H_2 by Ethanol Photoreforming on Au/TiO_2. *Adv. Funct. Mater.* **2014**, *24*, 241–248. [CrossRef]
9. Khatun, F.; Abd Aziz, A.; Sim, L.C.; Monir, M.U. Plasmonic Enhanced Au Decorated TiO_2 Nanotube Arrays as a Visible Light Active Catalyst towards Photocatalytic CO_2 Conversion to CH_4. *J. Environ. Chem. Eng.* **2019**, *7*, 103233. [CrossRef]
10. Galin, A.; Walendziewski, J. Photocatalytic Water Splitting over Pt-TiO_2 in the Presence of Sacrificial Reagents. *Energy Fuels* **2005**, *19*, 1143–1147.

1. Xiong, Z.; Wang, H.; Xu, N.; Li, H.; Fang, B.; Zhao, Y.; Zhang, J.; Zheng, C. Photocatalytic Reduction of CO_2 on Pt^{2+}-Pt0/TiO_2 Nanoparticles under UV/Vis Light Irradiation: A Combination of Pt^{2+} Doping and Pt Nanoparticles Deposition. *Int. J. Hydrogen Energy* **2015**, *40*, 10049–10062. [CrossRef]
2. Luna, A.L.; Dragoe, D.; Wang, K.; Beaunier, P.; Kowalska, E.; Ohtani, B.; Bahena Uribe, D.; Valenzuela, M.A.; Remita, H.; Colbeau-Justin, C. Photocatalytic Hydrogen Evolution Using Ni-Pd/TiO_2: Correlation of Light Absorption, Charge-Carrier Dynamics, and Quantum Efficiency. *J. Phys. Chem. C* **2017**, *121*, 14302–14311. [CrossRef]
3. Ong, W.-J.; Gui, M.M.; Chai, S.-P.; Mohamed, A.R. Direct Growth of Carbon Nanotubes on Ni/TiO_2 as next Generation Catalysts for Photoreduction of CO_2 to Methane by Water under Visible Light Irradiation. *RSC Adv.* **2013**, *3*, 4505. [CrossRef]
4. Bolbol, A.M.; Abd-Elkader, O.H.; Elshimy, H.; Zaki, Z.I.; Shata, S.A.; Kamel, M.; Radwan, A.S.; Mostafa, N.Y. The Effect of Zr (IV) Doping on TiO_2 Thin Film Structure and Optical Characteristics. *Results Phys.* **2022**, *42*, 105955. [CrossRef]
5. Juma, A.; Oja Acik, I.; Oluwabi, A.T.; Mere, A.; Mikli, V.; Danilson, M.; Krunks, M. Zirconium Doped TiO_2 Thin Films Deposited by Chemical Spray Pyrolysis. *Appl. Surf. Sci.* **2016**, *387*, 539–545. [CrossRef]
6. Oluwabi, A.T.; Juma, A.O.; Oja Acik, I.; Mere, A.; Krunks, M. Effect of Zr Doping on the Structural and Electrical Properties of Spray Deposited TiO_2 Thin Films. *Proc. Est. Acad. Sci.* **2018**, *67*, 147. [CrossRef]
7. Mbiri, A.; Taffa, D.H.; Gatebe, E.; Wark, M. Zirconium Doped Mesoporous TiO_2 Multilayer Thin Films: Influence of the Zirconium Content on the Photodegradation of Organic Pollutants. *Catal. Today* **2019**, *328*, 71–78. [CrossRef]
8. Taechasirivichai, K.; Chiarakorn, S.; Chawengkijwanich, C.; Pongprayoon, T.; Chuangchote, S. Ceramic Tiles Coated with Zr-Ag Co-doped TiO_2 Thin Film for Indoor Air Purifying and Antimicrobial Applications. *Int. J. Appl. Ceram. Technol.* **2022**, *20*, 2019–2029. [CrossRef]
9. Song, J.; Wang, X.; Yan, J.; Yu, J.; Sun, G.; Ding, B. Soft Zr-Doped TiO_2 Nanofibrous Membranes with Enhanced Photocatalytic Activity for Water Purification. *Sci. Rep.* **2017**, *7*, 1636. [CrossRef]
10. Schiller, R.; Weiss, C.K.; Landfester, K. Phase Stability and Photocatalytic Activity of Zr-Doped Anatase Synthesized in Miniemulsion. *Nanotechnology* **2010**, *21*, 405603. [CrossRef]
11. Bathula, C.; Nissimagoudar, A.S.; Kumar, S.; Jana, A.; Sekar, S.; Lee, S.; Kim, H.-S. Enhanced Photodegradation of Methylene Blue by Zr-Doped TiO_2:A Combined DFT and Experimental Investigation. *Inorg. Chem. Commun.* **2024**, *164*, 112442. [CrossRef]
12. Wang, C.; Geng, A.; Guo, Y.; Jiang, S.; Qu, X.; Li, L. A Novel Preparation of Three-Dimensionally Ordered Macroporous M/Ti (M=Zr or Ta) Mixed Oxide Nanoparticles with Enhanced Photocatalytic Activity. *J. Colloid Interface Sci.* **2006**, *301*, 236–247. [CrossRef] [PubMed]
13. Divya, G.; Jaishree, G.; Sivarao, T.; Lakshmi, K.V.D. Microwave Assisted Sol–Gel Approach for Zr Doped TiO_2 as a Benign Photocatalyst for Bismark Brown Red Dye Pollutant. *RSC Adv.* **2023**, *13*, 8692–8705. [CrossRef]
14. Liao, X.; Ren, H.-T.; Shen, B.; Lin, J.-H.; Lou, C.-W.; Li, T.-T. Enhancing Mechanical and Photocatalytic Properties by Surface Microstructure Regulation of TiO_2 Nanofiber Membranes. *Chemosphere* **2023**, *313*, 137195. [CrossRef] [PubMed]
15. Lukáč, J.; Klementová, M.; Bezdička, P.; Bakardjieva, S.; Šubrt, J.; Szatmáry, L.; Bastl, Z.; Jirkovský, J. Influence of Zr as TiO_2 Doping Ion on Photocatalytic Degradation of 4-Chlorophenol. *Appl. Catal. B* **2007**, *74*, 83–91. [CrossRef]
16. Chattopadhyay, S.; Mondal, S.; De, G. Hierarchical $Ti_{1-x}Zr_xO_{2-y}$ Nanocrystals with Exposed High Energy Facets Showing Co-Catalyst Free Solar Light Driven Water Splitting and Improved Light to Energy Conversion Efficiency. *J. Mater. Chem. A Mater.* **2017**, *5*, 17341–17351. [CrossRef]
17. Pahi, S.; Sahu, S.; Singh, S.K.; Behera, A.; Patel, R.K. Visible Light Active Zr- and N-Doped TiO_2 Coupled g-C_3N_4 Heterojunction Nanosheets as a Photocatalyst for the Degradation of Bromoxynil and Rh B along with the H_2 Evolution Process. *Nanoscale Adv.* **2021**, *3*, 6468–6481. [CrossRef]
18. Barba-Nieto, I.; Caudillo-Flores, U.; Gómez-Cerezo, M.N.; Kubacka, A.; Fernández-García, M. Boosting Pt/TiO_2 Hydrogen Photoproduction through Zr Doping of the Anatase Structure: A Spectroscopic and Mechanistic Study. *Chem. Eng. J.* **2020**, *398*, 125665. [CrossRef]
19. Pérez-Larios, A.; Rico, J.L.; Anaya-Esparza, L.M.; Vargas, O.A.G.; González-Silva, N.; Gómez, R. Hydrogen Production from Aqueous Methanol Solutions Using Ti–Zr Mixed Oxides as Photocatalysts under UV Irradiation. *Catalysts* **2019**, *9*, 938. [CrossRef]
20. Li, Y.; Xiang, Y.; Peng, S.; Wang, X.; Zhou, L. Modification of Zr-Doped Titania Nanotube Arrays by Urea Pyrolysis for Enhanced Visible-Light Photoelectrochemical H_2 Generation. *Electrochim. Acta* **2013**, *87*, 794–800. [CrossRef]
21. Touam, T.; Atoui, M.; Hadjoub, I.; Chelouche, A.; Boudine, B.; Fischer, A.; Boudrioua, A.; Doghmane, A. Effects of Dip-Coating Speed and Annealing Temperature on Structural, Morphological and Optical Properties of Sol-Gel Nano-Structured TiO_2 Thin Films. *Eur. Phys. J. Appl. Phys.* **2014**, *67*, 30302. [CrossRef]
22. Oh, S.H.; Kim, D.J.; Hahn, S.H.; Kim, E.J. Comparison of Optical and Photocatalytic Properties of TiO_2 Thin Films Prepared by Electron-Beam Evaporation and Sol–Gel Dip-Coating. *Mater. Lett.* **2003**, *57*, 4151–4155. [CrossRef]
23. Ranjitha, A.; Muthukumarasamy, N.; Thambidurai, M.; Balasundaraprabhu, R.; Agilan, S. Effect of Annealing Temperature on Nanocrystalline TiO_2 Thin Films Prepared by Sol–Gel Dip Coating Method. *Optik* **2013**, *124*, 6201–6204. [CrossRef]
24. Johari, N.D.; Rosli, Z.M.; Juoi, J.M.; Yazid, S.A. Comparison on the TiO_2 Crystalline Phases Deposited via Dip and Spin Coating Using Green Sol–Gel Route. *J. Mater. Res. Technol.* **2019**, *8*, 2350–2358. [CrossRef]
25. Teeparthi, S.R.; Awin, E.W.; Kumar, R. Dominating Role of Crystal Structure over Defect Chemistry in Black and White Zirconia on Visible Light Photocatalytic Activity. *Sci. Rep.* **2018**, *8*, 5541. [CrossRef] [PubMed]

36. Shaddad, M.N.; Cardenas-Morcoso, D.; García-Tecedor, M.; Fabregat-Santiago, F.; Bisquert, J.; Al-Mayouf, A.M.; Gimenez, S. TiO Nanotubes for Solar Water Splitting: Vacuum Annealing and Zr Doping Enhance Water Oxidation Kinetics. *ACS Omega* **2019**, 16095–16102. [CrossRef] [PubMed]
37. Godoy Junior, A.; Pereira, A.; Gomes, M.; Fraga, M.; Pessoa, R.; Leite, D.; Petraconi, G.; Nogueira, A.; Wender, H.; Miyakawa, W et al. Black TiO_2 Thin Films Production Using Hollow Cathode Hydrogen Plasma Treatment: Synthesis, Material Characteristic and Photocatalytic Activity. *Catalysts* **2020**, *10*, 282. [CrossRef]
38. Flores-Caballero, A.A.; Manzo-Robledo, A.; Alonso-Vante, N. The Cerium/Boron Insertion Impact in Anatase Nano-Structure on the Photo-Electrochemical and Photocatalytic Response. *Surfaces* **2021**, *4*, 54–65. [CrossRef]
39. Naumenko, A.; Gnatiuk, I.; Smirnova, N.; Eremenko, A. Characterization of Sol–Gel Derived TiO_2/ZrO_2 Films and Powders b Raman Spectroscopy. *Thin Solid Film.* **2012**, *520*, 4541–4546. [CrossRef]
40. Ji, C.H.; Kim, K.T.; Oh, S.Y. High-Detectivity Perovskite-Based Photodetector Using a Zr-Doped TiO_x Cathode Interlayer. *RSC Adv.* **2018**, *8*, 8302–8309. [CrossRef]
41. Bensouici, F.; Bououdina, M.; Dakhel, A.A.; Souier, T.; Tala-Ighil, R.; Toubane, M.; Iratni, A.; Liu, S.; Cai, W. Al Doping Effect on the Morphological, Structural and Photocatalytic Properties of TiO_2 Thin Layers. *Thin Solid Film.* **2016**, *616*, 655–661. [CrossRef]
42. Lee, S.-M.; Joo, Y.-H.; Kim, C.-I. Influences of Film Thickness and Annealing Temperature on Properties of Sol–Gel Derived $ZnO–SnO_2$ Nanocomposite Thin Film. *Appl. Surf. Sci.* **2014**, *320*, 494–501. [CrossRef]
43. Nakanishi, Y.; Kato, K.; Horikawa, M.; Yonekura, M. Influence of Zn-Sn Ratio on Optical Property and Microstructure of Zn-Sn-O Films Deposited by Magnetron Sputtering. *Thin Solid Film.* **2016**, *612*, 231–236. [CrossRef]
44. Sanchez-Martinez, A.; Koop-Santa, C.; Ceballos-Sanchez, O.; López-Mena, E.R.; González, M.A.; Rangel-Cobián, V.; Orozco Guareño, E.; García-Guaderrama, M. Study of the Preparation of TiO_2 Powder by Different Synthesis Methods. *Mater. Res Express* **2019**, *6*, 085085. [CrossRef]
45. Haryński, Ł.; Olejnik, A.; Grochowska, K.; Siuzdak, K. A Facile Method for Tauc Exponent and Corresponding Electronic Transitions Determination in Semiconductors Directly from UV–Vis Spectroscopy Data. *Opt. Mater.* **2022**, *127*, 112205. [CrossRef]
46. Schubert, E.F. *Physical Foundations of Solid-State Devices*, 2022nd ed.; E. Fred Schubert: New York, NY, USA, 2020.
47. Fang, D.; Huang, K.; Liu, S.; Huang, J. Fabrication and Photoluminiscent Properties of Titanium Oxide Nanotube Arrays. *J. Braz Chem. Soc.* **2008**, *19*, 1059–1064. [CrossRef]
48. Méndez-López, A.; Zelaya-Ángel, O.; Toledano-Ayala, M.; Torres-Pacheco, I.; Pérez-Robles, J.F.; Acosta-Silva, Y.J. The Influence of Annealing Temperature on the Structural and Optical Properties of ZrO_2 Thin Films and How Affects the Hydrophilicity. *Crystals* **2020**, *10*, 454. [CrossRef]
49. Nabi, G.; Raza, W.; Tahir, M.B. Green Synthesis of TiO_2 Nanoparticle Using Cinnamon Powder Extract and the Study of Optical Properties. *J. Inorg. Organomet. Polym. Mater.* **2020**, *30*, 1425–1429. [CrossRef]
50. Komaraiah, D.; Radha, E.; Sivakumar, J.; Ramana Reddy, M.V.; Sayanna, R. Photoluminescence and Photocatalytic Activity of Spin Coated Ag+ Doped Anatase TiO_2 Thin Films. *Opt. Mater.* **2020**, *108*, 110401. [CrossRef]
51. Hou, L.; Zhang, M.; Guan, Z.; Li, Q.; Yang, J. Effect of Annealing Ambience on the Formation of Surface/Bulk Oxygen Vacancies in TiO_2 for Photocatalytic Hydrogen Evolution. *Appl. Surf. Sci.* **2018**, *428*, 640–647. [CrossRef]
52. Goswami, P.; Ganguli, J.N. Tuning the Band Gap of Mesoporous Zr-Doped TiO_2 for Effective Degradation of Pesticide Quinalphos *Dalton Trans.* **2013**, *42*, 14480. [CrossRef] [PubMed]
53. Tudu, B.; Nalajala, N.; Saikia, P.; Gopinath, C.S. Cu–Ni Bimetal Integrated TiO_2 Thin Film for Enhanced Solar Hydrogen Generation. *Sol. RRL* **2020**, *4*, 1900557. [CrossRef]
54. Cruz, M.R.A.; Sanchez-Martinez, D.; Torres-Martínez, L.M. Optical Properties of TiO_2 Thin Films Deposited by DC Sputtering and Their Photocatalytic Performance in Photoinduced Process. *Int. J. Hydrogen Energy* **2019**, *44*, 20017–20028. [CrossRef]
55. Alfaro Cruz, M.R.; Sanchez-Martinez, D.; Torres-Martínez, L.M. TiO_2 Nanorods Grown by Hydrothermal Method and Their Photocatalytic Activity for Hydrogen Production. *Mater. Lett.* **2019**, *237*, 310–313. [CrossRef]
56. Zdorovets, M.; Kozlovskiy, A.; Tishkevich, D.; Zubar, T.; Trukhanov, A. The Effect of Doping of TiO_2 Thin Films with Low-Energy O^{2+} Ions on Increasing the Efficiency of Hydrogen Evolution in Photocatalytic Reactions of Water Splitting. *J. Mater. Sci. Mater. Electron.* **2020**, *31*, 21142–21153. [CrossRef]
57. Khan, M.A.; Sinatra, L.; Oufi, M.; Bakr, O.M.; Idriss, H. Evidence of Plasmonic Induced Photocatalytic Hydrogen Production on Pd/TiO_2 Upon Deposition on Thin Films of Gold. *Catal. Lett.* **2017**, *147*, 811–820. [CrossRef]
58. Naseri, N.; Kim, H.; Choi, W.; Moshfegh, A.Z. Optimal Ag Concentration for H_2 Production via $Ag:TiO_2$ Nanocomposite Thin Film Photoanode. *Int. J. Hydrogen Energy* **2012**, *37*, 3056–3065. [CrossRef]
59. Alenzi, N.; Liao, W.-S.; Cremer, P.S.; Sanchez-Torres, V.; Wood, T.K.; Ehlig-Economides, C.; Cheng, Z. Photoelectrochemical Hydrogen Production from Water/Methanol Decomposition Using Ag/TiO_2 Nanocomposite Thin Films. *Int. J. Hydrogen Energy* **2010**, *35*, 11768–11775. [CrossRef]
60. Dholam, R.; Patel, N.; Santini, A.; Miotello, A. Efficient Indium Tin Oxide/Cr-Doped-TiO_2 Multilayer Thin Films for H_2 Production by Photocatalytic Water-Splitting. *Int. J. Hydrogen Energy* **2010**, *35*, 9581–9590. [CrossRef]
61. Liu, E.; Kang, L.; Yang, Y.; Sun, T.; Hu, X.; Zhu, C.; Liu, H.; Wang, Q.; Li, X.; Fan, J. Plasmonic Ag Deposited TiO_2 Nano-Sheet Film for Enhanced Photocatalytic Hydrogen Production by Water Splitting. *Nanotechnology* **2014**, *25*, 165401. [CrossRef]
62. Strataki, N.; Lianos, P. Optimization of Parameters for Hydrogen Production by Photocatalytic Alcohol Reforming in the Presence of Pt/TiO_2 Nanocrystalline Thin Films. *J. Adv. Oxid. Technol.* **2008**, *11*, 111–115. [CrossRef]

63. Dholam, R.; Patel, N.; Adami, M.; Miotello, A. Physically and Chemically Synthesized TiO_2 Composite Thin Films for Hydrogen Production by Photocatalytic Water Splitting. *Int. J. Hydrogen Energy* **2008**, *33*, 6896–6903. [CrossRef]
64. Ibarra-Rodríguez, L.I.; Garay-Rodríguez, L.F.; Hernández-Majalca, B.C.; Domínguez-Arvizu, J.L.; López-Ortiz, A.; Torres-Martínez, L.M.; Collins-Martínez, V.H. N–TiO_2/MO (M: Ni, Cu) Films for Hydrogen Production Using Visible Light. *Int. J. Hydrogen Energy*, 2024; *in press*. [CrossRef]
65. Méndez, F.J.; Barrón-Romero, D.; Pérez, O.; Flores-Cruz, R.D.; Rojas-Challa, Y.; García-Macedo, J.A. A Highly Efficient and Recyclable Pt/TiO_2 Thin Film Photocatalytic System for Sustainable Hydrogen Production. *Mater. Chem. Phys.* **2023**, *305*, 127925. [CrossRef]

surfaces

Review

The Use of Magnetic Porous Carbon Nanocomposites for the Elimination of Organic Pollutants from Wastewater

Bendi Anjaneyulu [1], Ravi Rana [2], Versha [2,*], Mozhgan Afshari [3] and Sónia A. C. Carabineiro [4,*]

1 Department of Chemistry, Presidency University, Rajanukunte, Itgalpura, Bangalore 560064, Karnataka, India; anjaneyulu.bendi@gmail.com
2 Department of Chemistry, Faculty of Science, Baba Masthnath University, Rohtak 124001, Haryana, India; hit.ravirana@gmail.com
3 Department of Chemistry, Shoushtar Branch, Islamic Azad University, Shoushtar, Iran; m.afshari@iau-shoushtar.ac.ir
4 LAQV-REQUIMTE, Department of Chemistry, NOVA School of Science and Technology, Universidade NOVA de Lisboa, 2829-516 Caparica, Portugal
* Correspondence: versha2195@gmail.com (V.); sonia.carabineiro@fct.unl.pt (S.A.C.C.)

Abstract: One of the most significant challenges the world is currently facing is wastewater treatment. A substantial volume of effluents from diverse sources releases numerous pollutants into the water. Among these contaminants, organic pollutants are particularly concerning due to the associated risk of being released into the environment, garnering significant attention. Rapid advancements in agriculture and industry on a global scale generate vast volumes of hazardous organic compounds, which eventually find their way into natural systems. Recently, the release of industrial wastewater has been increasing, due to the progress of numerous businesses. This poses a danger to humans and the environment, leading to environmental contamination. The application of carbon nanocomposites in applied nanotechnology has recently expanded due to their large surface area, substantial pore volume, low preparation cost, and environmental resilience. Expanding the use of nanomaterials in water treatment is essential, as magnetic carbon nanocomposites consistently demonstrate an efficient elimination of pollutants from water solutions. In the current study, we have highlighted the application of magnetic porous carbon nanocomposites in removing organic pollutants from wastewater.

Keywords: magnetic carbon nanocomposites; pollutants; adsorption; wastewater

1. Introduction

Wastewater reclamation or reuse has become an imperative need in the current scenario, given the rapid depletion of freshwater supplies. Agriculture alone accounts for 92% of global water usage [1–5], with approximately 70% of freshwater from rivers and subsurface sources dedicated to irrigation. This alarming statistic raises serious concerns for nations struggling with water scarcity [6–9]. Therefore, incorporating wastewater recycling in agriculture emerges as a pivotal strategy to substitute the use of freshwater [10].

In general, treated wastewater finds its application in various non-potable uses, including firefighting, vehicle washing, toilet flushing, irrigation, groundwater replenishment, golf course irrigation, building construction [11–16], and even for cooling purposes in thermal power plants. Both domestic water and treated wastewater contain a range of nutrients, such as nitrogen, phosphorus, sulfur and potassium. Notably, a significant portion of the nitrogen and phosphorus in wastewater is readily absorbable by plants, making it a viable option for irrigation [17,18].

Currently, diverse conventional technologies for wastewater treatment have been devised, including membrane filtration [19,20], flocculation [21,22], coagulation [23], Fenton reagent [24,25], electrolysis [26] and photocatalytic oxidation [27]. Among the various available technologies, the adsorption method is acknowledged as a more cost-effective, efficient, practical and environmentally friendly alternative for wastewater treatment [28–30]. Various adsorbents have been examined for their effectiveness in purifying wastewater, including zeolites [31–33], clay minerals [34,35], kaolinite [36–38], pillared clays [39,40], silica gel [41,42], activated carbon [43–45] and biomass [46–48]. However, many existing adsorbents exhibit limited efficiency in removing specific contaminants in water, such as heavy metals and oil [49,50]. Numerous studies have emphasized the capability of activated carbon as an important adsorbent for the treatment of wastewater, owing to its significant porosity and expansive surface area [51–53].

The increased focus on nanotechnology has sparked a notable surge in the application of nanostructured materials across various domains, particularly in environmental remediation and wastewater treatment [54–56]. Nanostructured materials have particle sizes below 100 nm. They appear in diverse forms, including nanoparticles, nanotubes, nanowires and nanofibers [57–59]. Nanomaterials demonstrate superior adsorption abilities for various water contaminants when compared to larger bulk materials [60–62].

Carbon-based nanomaterials (CNMs) have garnered substantial attention as cutting-edge materials across various applications within the spectrum of nanostructured materials. This increased attention is due to their remarkable physical and chemical traits, alongside outstanding thermal, mechanical and electrical conductive characteristics [63–65]. The remarkable attributes of CNMs represent a significant advancement in technology, finding applications in diverse fields such as biosensors [66,67], reinforcing composite materials [68,69], materials with electrical conductivity [70], drug delivery [71,72], the biomedical field [73] and catalyst support [74,75]. Additionally, CNMs are widely acknowledged for their efficacy as superior adsorbents in wastewater pollutant removal [76–78].

Recently, there has been a growing interest in magnetic nanoparticles (NPs) in various environmental engineering applications. Spanning sizes between 1 and 100 nm and featuring a notable ratio of surface area to volume, with a capacity for carrying a high load, these nanoparticles have shown to be efficient materials used to trap pollutants from polluted water [79,80].

Despite the efficient adsorption capacity exhibited by carbon nanomaterials, their practical use faces challenges due to their strong adaptability and hydrophilic nature, complicating their retrieval and separation from water [81]. Separation using magnets emerges as a swift and efficient post-adsorption technique, surpassing alternatives like filtration and centrifugation [82]. Consequently, using Fe oxide nanoparticles for adsorption has attracted considerable interest. Besides their adsorption ability, these nanomaterials are cost effective, abundant and environmentally friendly. However, their inherent reactivity, especially in nanoparticle form, necessitates stability achieved through surface coating [83].

Recent findings reveal that it is possible to enhance the removal of organic dyes by coating magnetic iron oxide nanoparticles with carbon materials, facilitating easy recovery and recycling due to the synergistic impact of Fe oxides and carbon nanomaterials [84]. These altered materials have shown effectiveness in eliminating a range of dyes, like Rhodamine B (RhB) [85], methylene blue (MB), food yellow, and various others.

This review offers a comprehensive summary of magnetic porous carbon nanocomposites (magnetic fullerene, magnetic carbon dots, magnetic carbon nanotubes, magnetic graphite, magnetic graphene, and magnetic graphene oxide) for eliminating organic pollutants from wastewater. This data will contribute to advancing the rational design of magnetic porous carbon nanocomposites for water treatment, thereby contributing to the development of improved technologies for ensuring adequate water quality. Magnetic porous carbon nanocomposites can also be used for the removal of heavy metals. The interested reader should consult relevant reviews on the subject [86,87].

2. The Application of Magnetic Porous Carbon Nanocomposites

2.1. Magnetic Fullerene Nanocomposites

Elessawy et al. obtained magnetic fullerene nanocomposites with functionalization (FMFNC) in a simple, one-step and environmentally friendly manner. The method involved utilizing the catalytic thermal breakdown of waste PET bottles as a starting material and employing ferrocene both as a catalyst and as a source of magnetic nanoparticles in combination for eliminating MB and acid blue 25 (AB25) dyes through an adsorption process (Scheme 1). Subsequently, the composite could be effortlessly isolated using an external magnet [88].

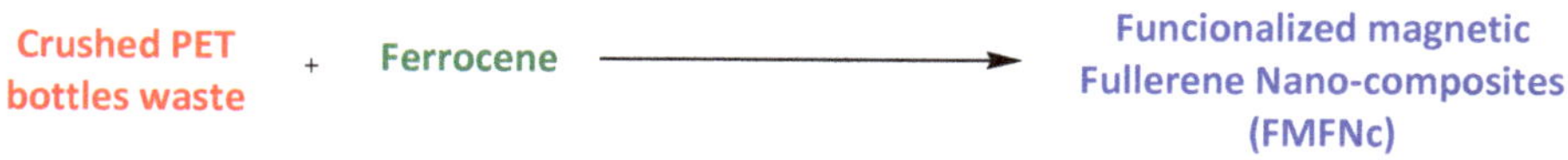

Scheme 1. Synthesis of FMFNc.

After 10 consecutive cycles, FMFNC showed impressive effectiveness in adsorbing MB or AB25. The adsorption mechanism was verified through an examination of surface chemistry both prior to and following adsorption. It was primarily achieved through hydrogen bonding, π-π stacking contact, and electrostatic interaction. The advantages of high adsorption capacity and convenient separation are combined when fullerene and magnetic nanoparticles are used (Schemes 2 and 3).

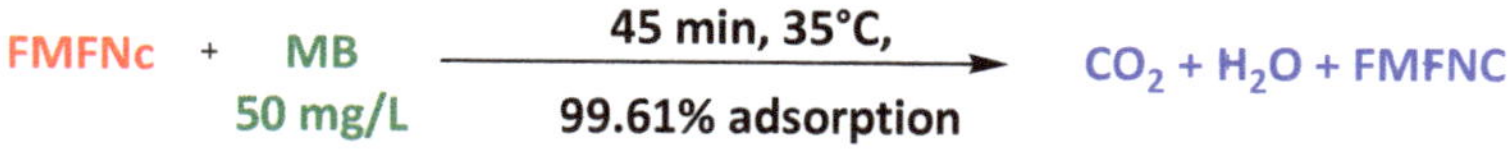

Scheme 2. Removal of methylene blue.

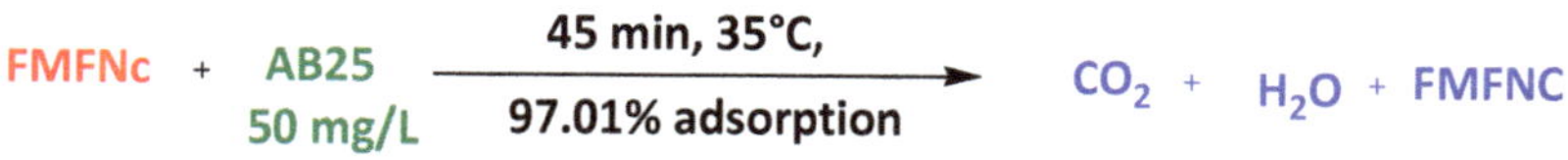

Scheme 3. Removal of acid blue 25.

Elessawy et al. obtained a new, simple, single-step method employed to produce functionalized magnetic fullerene nanocomposites (FMFN) by utilizing ferrocene both as a catalyst and a precursor for magnetite, along with the thermal catalytic decomposition of discarded poly(ethylene terephthalate) bottles as the source material [89] (Schemes 4 and 5).

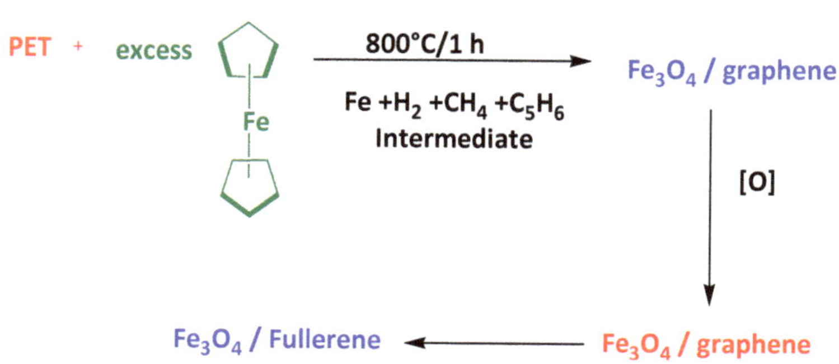

Scheme 4. Synthesis of fullerene.

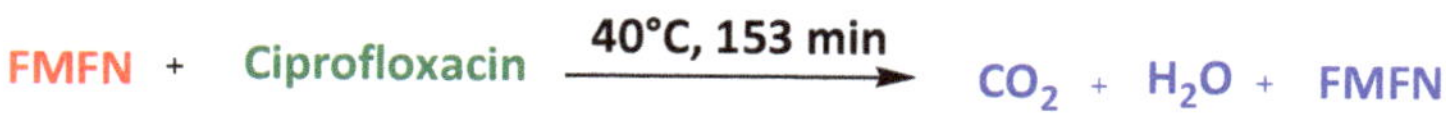

Scheme 5. Removal of ciprofloxacin.

According to the resolved equation, achieving the complete removal of ciprofloxacin requires optimal conditions: a contact time of 153 min and a ciprofloxacin amount of 65 mg/L at 40 °C. The spontaneity, exothermicity and increased casualness of the ciprofloxacin adsorption are confirmed by the negative ΔG and ΔH values, along with a positive ΔS value.

2.2. *Magnetic Carbon-Dot Nanocomposites*

Deng et al. developed a procedure for creating carbon nanotubes (CNTs) modified with carbon dots (CMNTs) in three sequential steps: firstly, preparing magnetic carbon nanotubes; secondly, synthesizing carbon dots; and lastly, modifying the magnetic multiwalled carbon nanotubes surfaces using these carbon dots for the removal of carbamazepine [90] (Scheme 6).

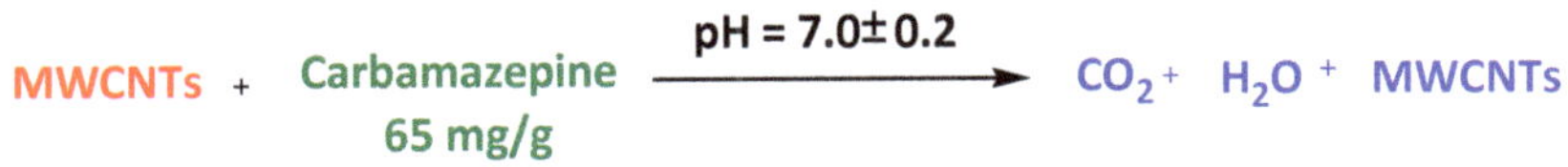

Scheme 6. Removal of carbamazepine.

At a pH of 7.0 ± 0.2, CMNTs showed a substantial adsorption of 65 mg/g, surpassing that of numerous previously documented adsorbents. The removal of carbamazepine by CMNTs showed swift removal within the initial 3 h, with kinetics effectively modeled by the HSDM and pseudo-second-order models. The simplified HSDM model effectively characterizes the declining adsorption capacity of CMNTs over time and illustrates the movement of carbamazepine particles within the CMNTs. The material used for adsorption can undergo regeneration and be reused repeatedly, experiencing a capacity reduction of less than 2.2% across six cycles.

Sun created magnetic carbon dots by combining C-dots with magnetic Fe_3O_4 nanoparticles. This synthesis aimed to investigate their photocatalytic performance in the presence of visible light and assess their recyclability in wastewater treatment. The specimens were created using a bottom-up method at a reaction temperature (*Tr*) set at 140 °C and 180 °C, with various reaction times (*tr* = 0–18 h). The outcomes revealed a progressive attachment of C–dots to Fe_3O_4 nanoparticles as *tr* increased at *Tr* = 140 °C, while a sudden move towards the maximum adsorption on Fe_3O_4 occurred at *Tr* = 180 °C [91] (Schemes 7 and 8).

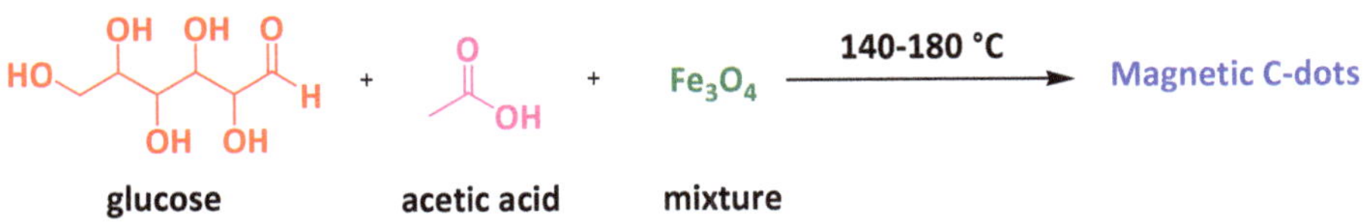

Scheme 7. Synthesis of magnetic C-dots.

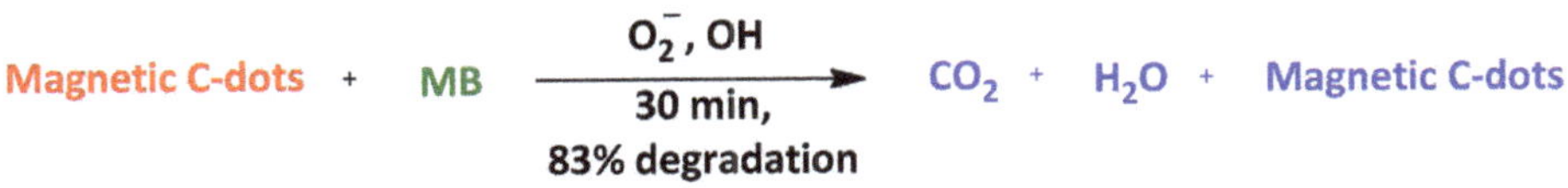

Scheme 8. Removal of methylene blue.

In just half an hour of exposure to visible light, the MB amount can decrease by 83%. The recyclability experiment distinctly indicated that the magnetic C-dots exhibited the capability for over a tenfold increase in photocatalytic degradation.

2.3. *Magnetic Carbon Nanotube Nanocomposites*

Zhang et al. developed a remarkably effective Fenton-like catalyst for breaking down organic pollutants by synthesizing MIL–88B–Fe with integrated CNTs. The latter possess surface oxygen functional groups rich in electrons. CNTs were integrated into the metal-organic framework to enhance Fe(II) content, thereby improving Fenton-like behavior. The synthesized CNT@MIL–88B–Fe (C@M) exhibited significantly enhanced catalytic capabilities.

The rate constant for the pseudo-first-order kinetics of phenol elimination using C@M (0.32 min^{-1}) surpassed that of MIL–88B–Fe by approximately sevenfold and rivaled or exceeded values reported for other heterogeneous Fenton-like materials. Furthermore, the Fenton-like system demonstrated an effective degradation of diverse persistent organic contaminants and showcased remarkable catalytic activity across a broad pH spectrum (4–9). XPS verified a gradual increase in the Fe(II) amount of the catalyst with CNT loadings [92] (Scheme 9).

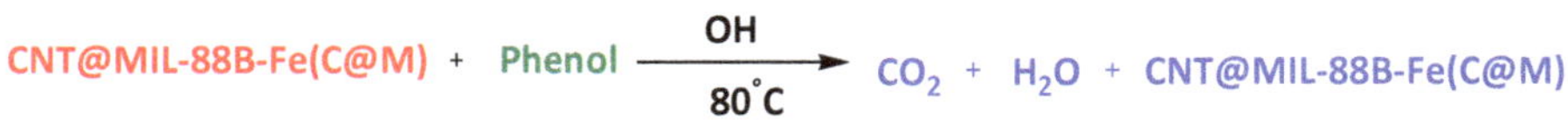

Scheme 9. Removal of phenol.

Zhang et al. successfully synthesized an adsorbent of magnetic CNTs functionalized with polyethyleneimine (PEI@MCNTs) and thoroughly examined its adsorption capacity to eliminate Alizarin Red S (ARS) from the dyeing effluent. PEI@MCNTs were made using the co-precipitation technique.

ARS could be effectively extracted from an acidic aqueous solution (pH $\leq$ 6.0) using PEI@MCNTs for 40 min at an ambient temperature. Thanks to their numerous contacts and abundance of adsorption sites, PEI@MCNTs exhibited a spontaneous adsorption process and good selectivity towards ARS. PEI@MCNTs had the highest adsorption capacity of 196.08 mg g^{-1} for ARS, as obtained by the Langmuir isotherm. The efficiency of removing ARS in tap, river and lake waters stayed within the range of 94.6%, 89.3%, and 91.8%, respectively, closely resembling the efficiency observed in distilled water (96.7%). This capacity exceeded that of conventional adsorbents currently on the market. Additionally, the PEI@MCNTs were easily regenerated using a 10 mM NaOH solution after being collected using an external magnet. The PEI@MCNTs prepared exhibit promise as adsorbents for efficiently removing anthraquinone dyes in extensive wastewater treatment [93] (Scheme 10).

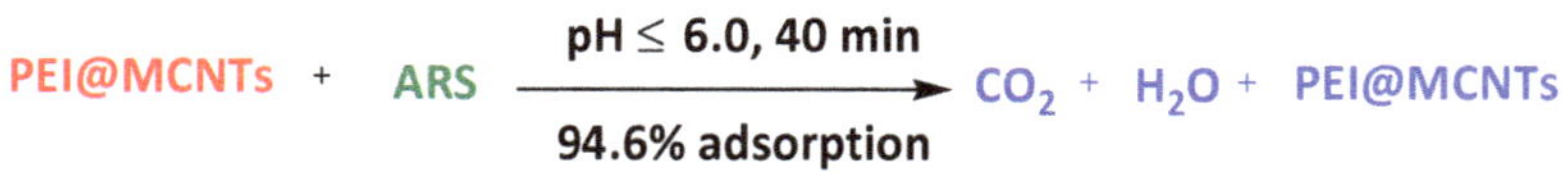

Scheme 10. Removal of Alizarin Red S.

Salam et al. developed a composite containing multi-walled carbon nanotubes (MWCNTs), magnetite and chitin, creating a magnetic nanocomposite with the purpose of efficiently eliminating Rose Bengal from both authentic and simulated solutions [94]. This magnetic nanocomposite was formed by physically combining MWCNTs with the biopolymer chitin and magnetite.

Various factors influencing the removal of Rose Bengal using an MCM nanocomposite were explored. The findings indicated that increasing the mass of MCM improved the adsorption process, with the optimal removal of Rose Bengal being achieved using 2 mg of MCM. Furthermore, the impact of adsorption time was investigated, revealing that equilibrium in adsorption was obtained after 30 min. The findings suggested that adsorption adhered to the pseudo-second-order kinetic model.

The outcomes revealed that the extraction of Rose Bengal from a water solution by an MCM composite was physically driven and endothermic. This was clear from the increased adsorption capability with rising solution temperature. Moreover, the negative ΔG, along with the positive ΔH and ΔS values, indicate that the adsorption of Rose Bengal is driven by entropy. When applied to remove Rose Bengal from an artificially contaminated wastewater sample, the MCM nanocomposite exhibited nearly 100% removal and retained its efficiency over five consecutive cycles (Scheme 11).

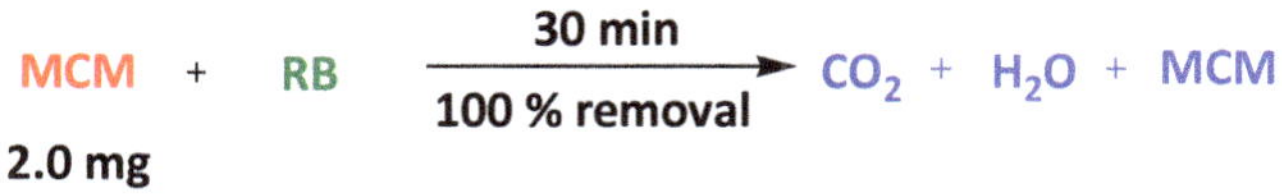

Scheme 11. Removal of Rose Bengal.

Cheng et al. developed a method to create composites of β-cyclodextrin (CD) attached to CNTs using cyclodextrins (CDs) and conducted reduction with the hydrazine hydrate of oxidized CNT. The resulting reduced samples (RCNT–CD) were affixed to Fe oxide during the creation of magnetic analogs (M–RCNT–CD).

M–RCNT–CD demonstrated a maximal adsorption of 196.5 mg/g of MB. The negative change in $\Delta G°$ and the positive change in $\Delta H°$ indicate that adsorption is spontaneous and endothermic, respectively; thus, the magnetic adsorbent is efficient for the removal of organic contaminants from water solutions [95] (Scheme 12).

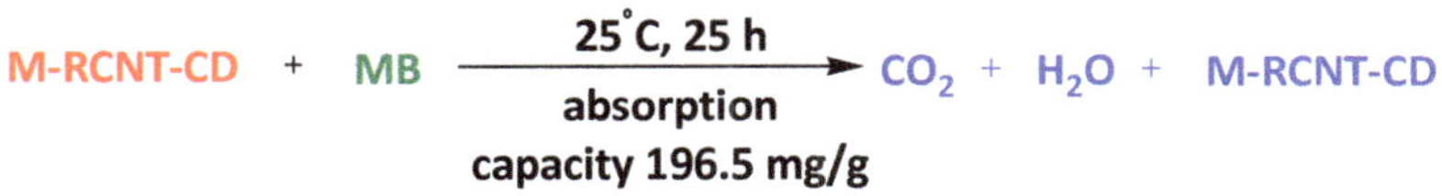

Scheme 12. Removal of methylene blue.

Zhu et al. synthesized magnetic graphitized MWCNTs modified with chitosan (CS–m–GMCNTs) using a cross-linking suspension method. The structure, form and magnetic characteristics of the synthesized CS–m–GMCNTs were assessed through several characterization techniques.

The model representing second-order kinetics, exhibiting high correlation coefficients ($R^2 > 0.998$), proved to be appropriate for characterizing the rate at which crystal violet is adsorbed onto CS–m–GMCNTs. A color removal of up to 94.58% and 100% was attainable within 100 min for water solutions of 10 and 60 mg L^{-1}, respectively.

Magnetic tests showed that CS–m–GMCNTs could attain a saturated magnetization of 12.27 emu g^{-1}. Additionally, the adsorption capability of crystal violet on CS–m–GMCNTs reached 263 mg g^{-1}. The thermodynamic parameter values suggested the adsorption had a significant reliance on the temperature of the liquid phase, suggesting a spontaneous and heat-releasing reaction.

Hence, CS–m–GMCNTs exhibit notable advantages, including very good dispersion in water, convenient separation and elevated adsorption capability. These characteristics suggest the possibility of its application in the efficient elimination of other carcinogenic and hazardous contaminants from water [96] (Scheme 13).

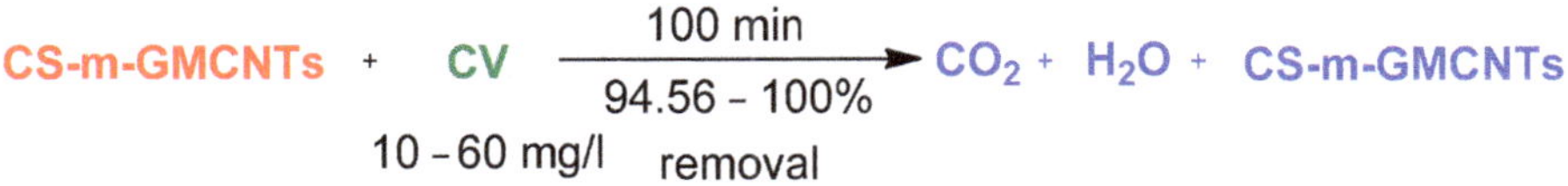

Scheme 13. Removal of crystal violet.

Gao et al. made a nanocomposite of magnetic polymer and MWCNTs to adsorb anionic azo dyes from water solutions.

The capturing capability of the magnetic polymer's multiwall carbon nanotube nanocomposite (MPMWCNT) was evaluated in solutions containing orange II, sunset yellow FCF and amaranth, and proved to be higher than that of magnetic MWCNTs without PGMIC. The adsorption capability sequence for anionic azo dyes was amaranth < orange II < sunset yellow

FCF. As the pH increased, the adsorption progressively diminished. The highest q_m values at 25 °C were 67.6, 85.5 and 47.4 mg g^{-1} for orange II, sunset yellow FCF and amaranth, respectively [97] (Schemes 14–16).

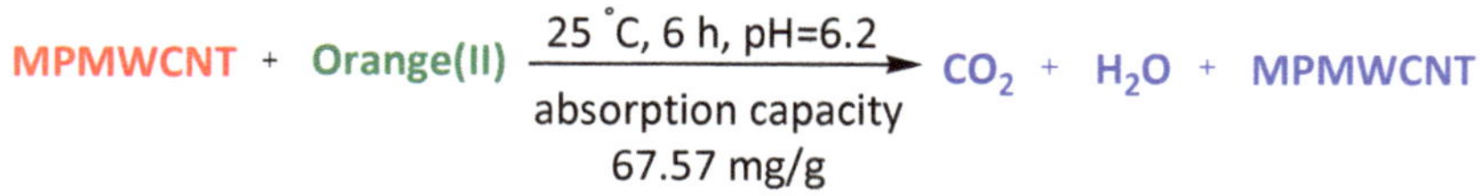

Scheme 14. Removal of orange (II).

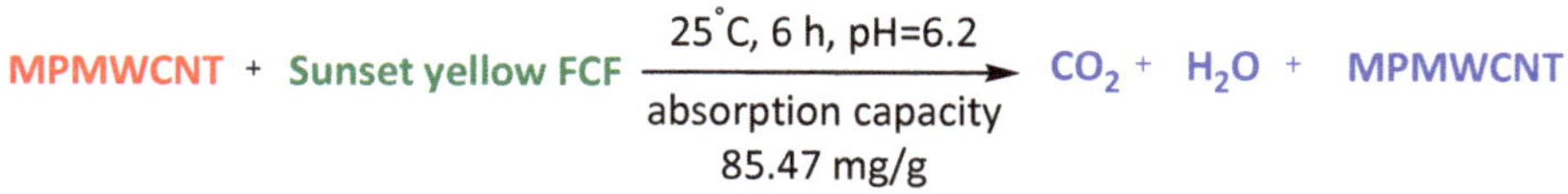

Scheme 15. Removal of sunset yellow FCF.

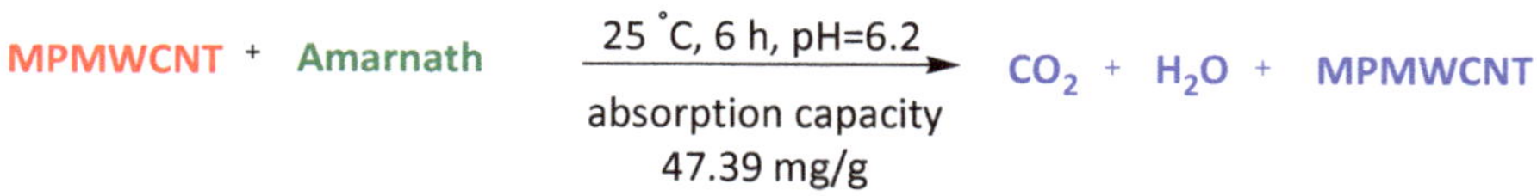

Scheme 16. Removal of amaranth.

Madrakian et al. focus on removing specific positively charged dyes from water using MWCNTs modified with magnets. MWCNTs loaded with Fe_3O_4 nanoparticles were synthesized through a straightforward solution-based approach. This involved the dissolving of ammonium iron (II) sulfate hexahydrate in water and a hydrazine hydrate solution (3:1 volume ratio), followed by the addition of pre-treated MWCNTs.

The best pH for eliminating all studied cationic dyes from water was 7.0. The highest anticipated adsorption amounts of CV, JG, Th and MB dyes were 228, 250, 36 and 48 mg g^{-1}, respectively. The desorption was additionally examined, employing acetonitrile as a solvent.

The variation in adsorption capability can be ascribed to distinctions in dye structures. JG and CV feature quaternary ammonium groups, while MB and Th are characterized as cationic sulfide dyes. The positive charge of JG and CV is distributed throughout the molecule, whereas in MB and Th, the positive charge accumulates around the heteroatom ring. Consequently, the adsorption of JG and CV onto the adsorbent proves more effective than that of MB and Th dyes. Maximum adsorption was achieved with 0.015 g of the adsorbent, resulting in approximately 95% removal for Th and MD and complete removal (100%) for JG and CV [98] (Schemes 17–20).

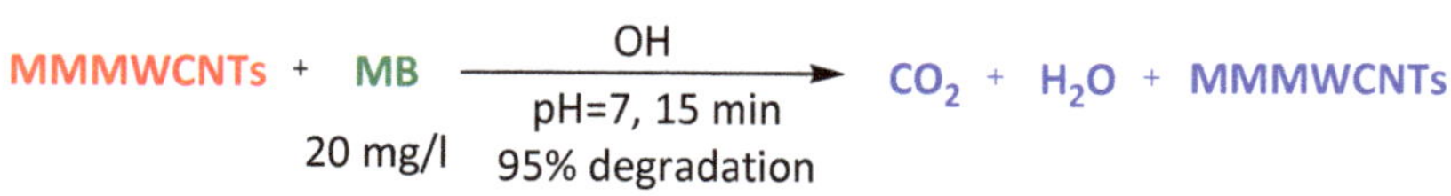

Scheme 17. Degradation of methylene blue.

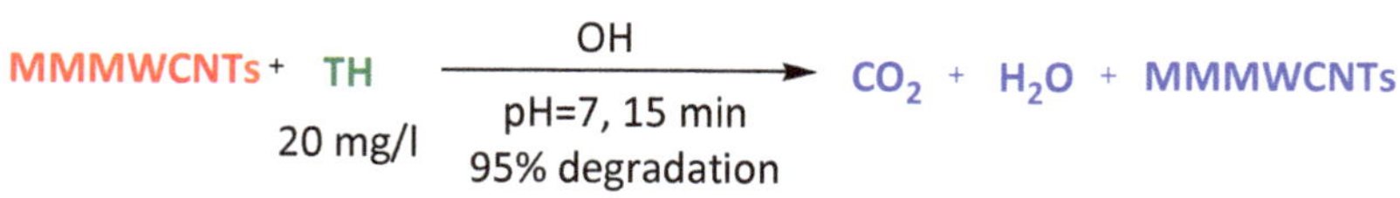

Scheme 18. Degradation of Thioflavin.

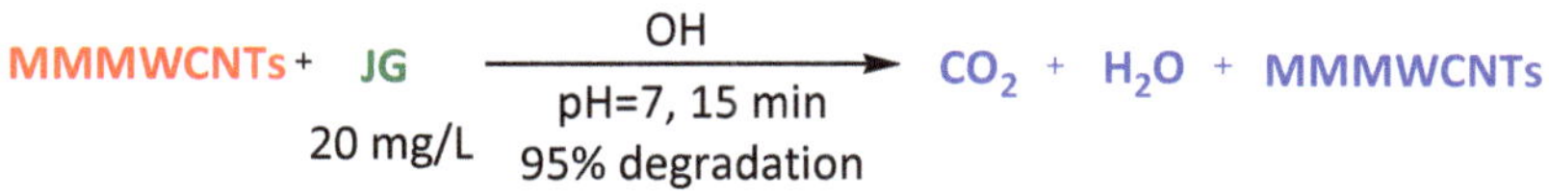

Scheme 19. Degradation of Janus Green.

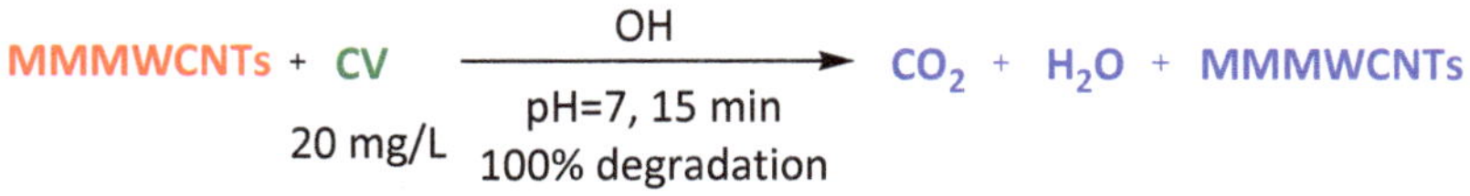

Scheme 20. Degradation of crystal violet.

Yan et al. created magnetic carbon nanotubes modified with guar gum for the purification of wastewater. Magnetic GG–MWCNT–Fe_3O_4 was synthesized by suspending GG–MWCNT in a solution containing $FeCl_3$·and $FeSO_4$ at 60 °C under a N_2 atmosphere. A solution of NH_3 was added, adjusting the mixture's pH between 10 and 11.

The isotherms showed that adsorption followed the Langmuir model, with GG–MWCNT–Fe_3O_4 exhibiting maximal adsorption of 61.9 mg g^{-1} for MB and 89.9 mg g^{-1} for NR. The magnetic GG–MWCNT–Fe_3O_4 demonstrates magnetic separation and adsorption capacity properties, making it a viable choice for eliminating pollutants from aqueous solutions [99] (Schemes 21 and 22).

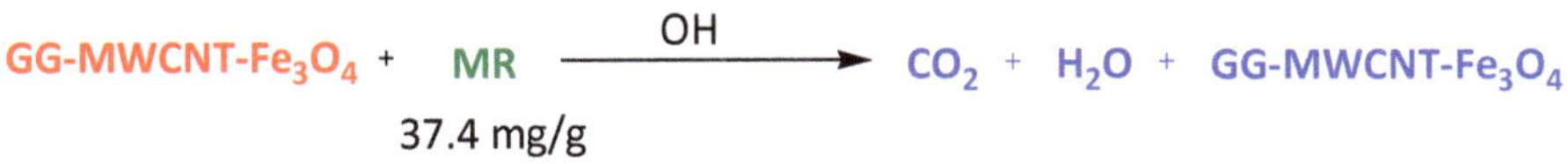

Scheme 21. Degradation of methylene blue.

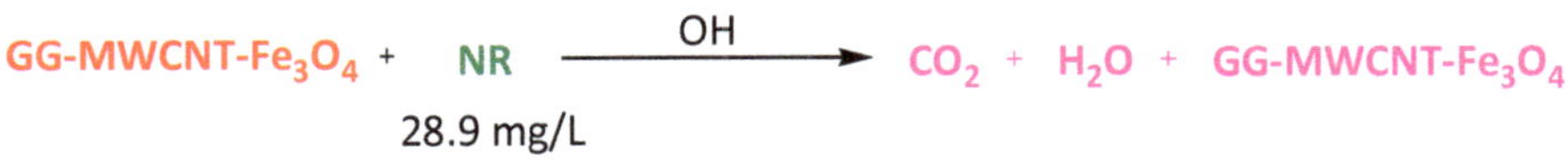

Scheme 22. Degradation of methylene blue.

Qu et al. fabricated MWCNTs filled with Fe_2O_3 particles to eliminate organic dyes from contaminated water. MWCNTs loaded with γ-Fe_2O_3 nanoparticles were synthesized through a hydrothermal reaction involving MWCNTs in a ferric nitrate solution, followed by calcination.

The dye adsorption test, involving neutral red (NR) and MB, reveals that equilibrium is reached within just 60 min. At pH 6, the adsorption capacities within the tested concentration span are 77.5 mg/g for NR and 42.3 mg/g for MB, making magnetic MWCNTs function as outstanding absorbents for these dyes. Unlike other materials, magnetic nanotubes not only demonstrate superior efficacy in dye adsorption but also offer convenient manipulation through an external magnetic field [100] (Schemes 23 and 24).

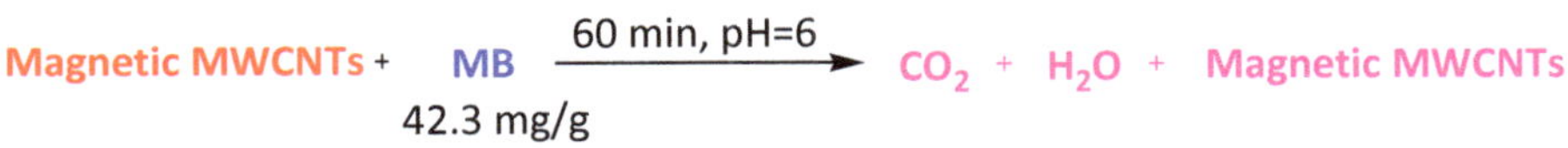

Scheme 23. Removal of methylene blue.

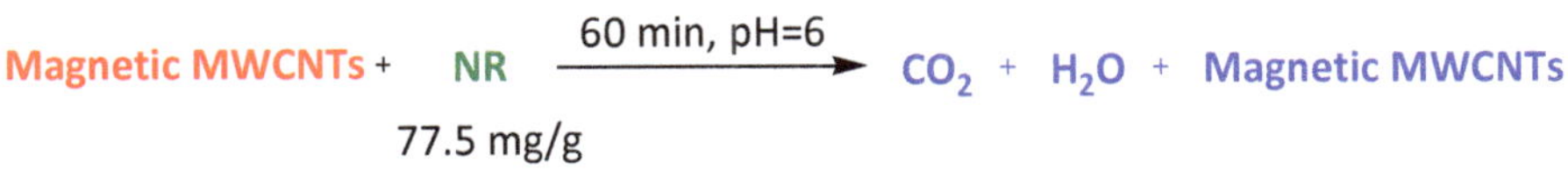

Scheme 24. Removal of neutral red.

2.4. Magnetic Graphite Nanocomposites

Ranjbar et al. synthesized magnetic graphite intercalation components as persulfate activators for Bisphenol A elimination from wastewater at pH 7. The "molten salt method" was employed to manufacture $FeCl_3$–GIC followed by calcination to generate magnetic GIC [101] (Scheme 25).

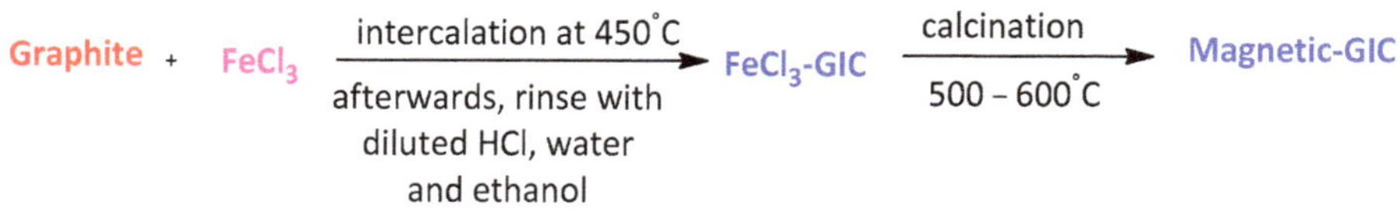

Scheme 25. Synthesis of magnetic– GIC.

With 2 g/L of catalyst and 1.2 g/L of persulfate at pH 7, the suggested method eliminated 99.3% of BPA in 75 min. Quenching assays revealed that the catalytic activity and removal of BPA were boosted by the production of free radicals and singlet oxygen, occurring through radical and non-radical routes. Additionally, the potential of recycling the manufactured catalyst was explored by its application to treated city wastewater. The findings indicated that the catalyst was capable of breaking down BPA in the wastewater in successive cycles, indicating its practicality (Scheme 26).

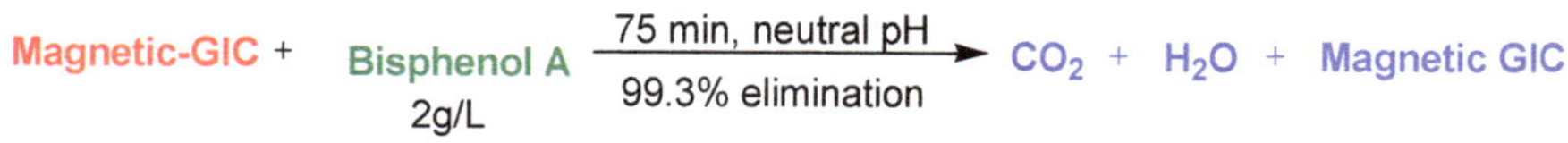

Scheme 26. Removal of Bisphenol A.

Ruan C. P et al. developed a Co/C magnetic nanocomposite resistant to acid by employing ZIF-67 as a template and precursor [102]. It underwent carbonization in the Ar atmosphere, followed by acid treatment. Through calcination in an inert atmosphere, Co nanoparticles formed and were evenly enveloped by graphite layers, facilitated by the catalytic effect of the Co-induced graphitization of carbon. These graphite layers protected against oxidation and acidic environments for the Co particles. Consequently, an acid-resistant magnetic adsorbent was created, suitable for application in a broad pH range (pH 1–13).

The magnetic Co/C, produced through synthesis, exhibited outstanding adsorption performance for two common dyes (RhB and malachite green, MG) across a broad pH spectrum, achieving a removal efficiency exceeding 99%. The adsorption behavior of these dyes was accurately explained by the Langmuir model in the adsorption isotherms. Notably, the maximum adsorption capacities for RhB and MG were 400.0 and 561.8 mg g^{-1}, respectively, surpassing the capacities of several mentioned adsorbents. Additionally, the adsorbent could be effortlessly regenerated through washing with ethylene glycol (EG), indicating remarkable reusability. After undergoing five reuse cycles, there was no observable decline in capacity. Moreover, the magnetic adsorbent demonstrated its practical use by attaining a removal efficiency surpassing 97% in eliminating organic dyes from household wastewater (Schemes 27 and 28).

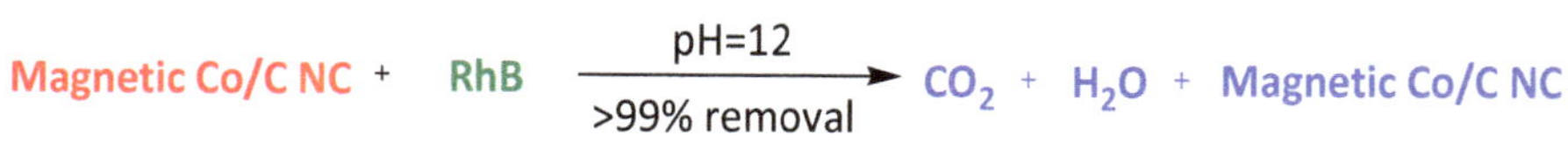

Scheme 27. Removal of Rhodamine B.

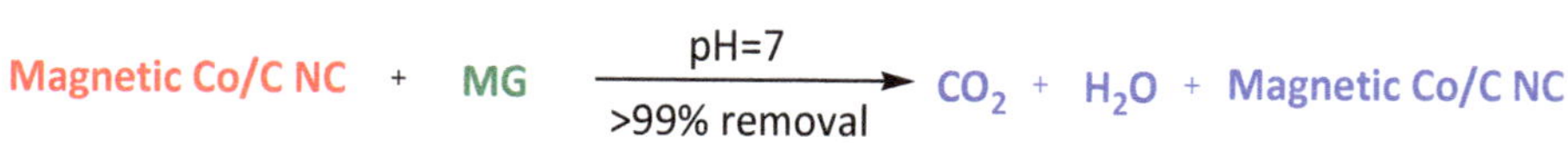

Scheme 28. Removal of malachite green.

Wang et al. produced a Fe_3O_4–graphite composite, which was synthesized, characterized and explored as a Fenton-like heterogeneous catalyst for levofloxacin (LEV) degradation in water. The composite was produced through a solvothermal one-step method and displayed remarkable characteristics for the elimination of LEV, accomplishing almost full elimination of 50 mg L^{-1} LEV within 15 min and 48% removal of total organic carbon in 60 min. The extensive conjugation electronic structure present in graphite might facilitate the rapid generation of $OH^{\bullet}$ radicals due to the facile reduction of Fe^{3+} to Fe^{2+}. Furthermore, it was noted that graphite could degrade LEV in H_2O_2. Hence, the combined effects of the graphitic structure and Fe_3O_4 MNPs likely promote the elevated catalytic activity of the composite. The efficiency of LEV degradation remained at ~80% during the fifth recycle, highlighting the potential applications of the material in water treatment for the removal of organic pollutants [103] (Scheme 29).

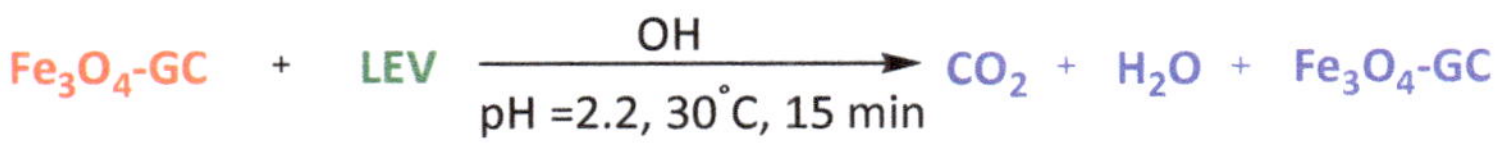

Scheme 29. Degradation of levofloxacin.

2.5. *Magnetic Graphene Nanocomposites*

Bharath et al. conducted studies focused on creating magnetite on porous graphene-based nanocomposites for novel adsorption and electrosorption techniques aimed at removing organic pollutants from wastewater. Initially, porous graphene was dispersed in water using ultrasonication. The resulting supernatant was collected. Subsequently, solutions of $FeCl_2$ and $FeCl_3$ were added, maintaining a pH range of 10–11 by introducing an ammonium hydroxide solution. After being kept at 180 °C for 12 h, it was washed with ethanol [104] (Scheme 30).

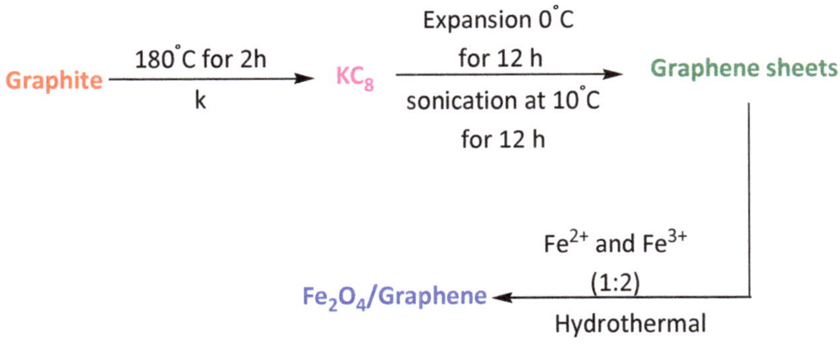

Scheme 30. Synthesis of Fe_2O_4/porous graphene nanocomposites.

The adsorption capacities of dyes were notably high in the case of Fe_3O_4/porous graphene nanocomposites. The adsorption of Fe_3O_4 on porous graphene was tested using methyl violet as an adsorbate. Fe_3O_4/porous graphene demonstrated swift adsorption within 5 min, possessing a high adsorption capability (Q_o–60 mg/g) and facilitating convenient separation and recyclability. This is ascribed to the elevated surface area resulting from the porosity of graphene and the strong magnetic properties of Fe_3O_4 (Scheme 31).

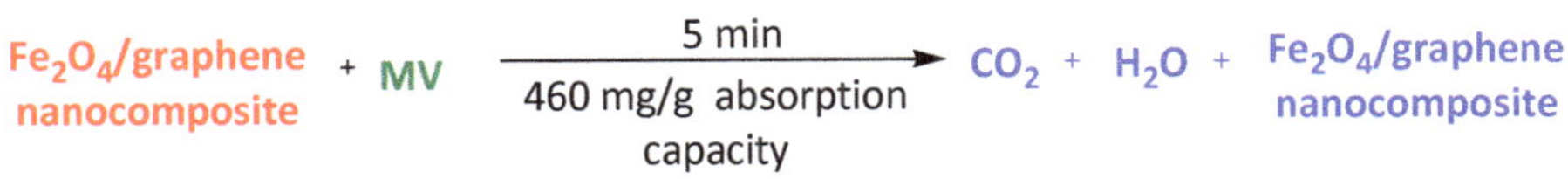

Scheme 31. Removal of methyl violet.

Yu et al. developed a magnetic sponge of graphene (Fe_3O_4–GS) designed for the elimination of MB. Fe_3O_4–GS was synthesized through lyophilization to facilitate the adsorption of dyes. Fe_3O_4–GS showed impressive adsorption of 526 mg/g for MB, exceeding the values reported for magnetic carbon nano-adsorbents in the existing literature. The kinetics of the adsorption of MB on Fe_3O_4–GS displayed a rapid rate, amenable to analysis through the pseudo-second-order and intraparticle diffusion models. The thermodynamic parameters indicate that, when feasible, the adsorption of MB on Fe_3O_4–GS should be conducted at elevated temperatures.

The thermodynamic investigation unveiled that the adsorption process was motivated by enhanced interface randomness. Additionally, both pH and ionic strength significantly impacted the adsorption capability of Fe_3O_4–GS [105] (Scheme 32).

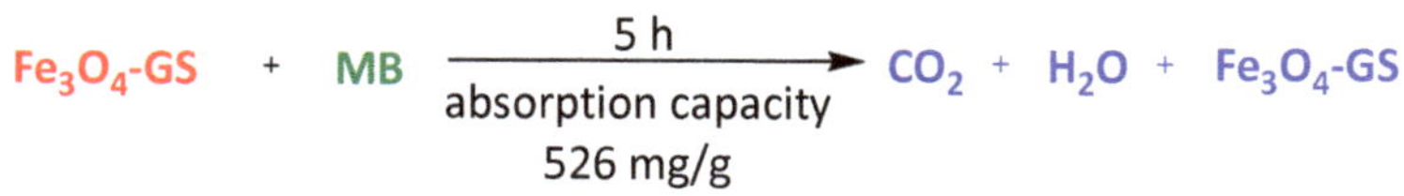

Scheme 32. Removal of methyl blue.

Zhao et al. synthesized a novel magnetic photocatalyst of graphene with Fe_3O_4 and NiO (GNs/Fe_3O_4/NiO), through a straightforward method. Three types of nanocomp–sites were produced with varying NiO contents: 50% (S1), 67% (S2) and 75% (S3), by weight.

The resulting mixture of GNs/Fe_3O_4, NiO MNPs and dodecyl benzene sulfonic acid sodium (SDBS) were sonicated in ethylene glycol and stirred at 25 °C in N_2 for 24 h, after which the product was isolated from the solution using a magnetic field. After washing with ethanol, GNs/Fe_3O_4/NiO nanocomposites were obtained [106] (Scheme 33).

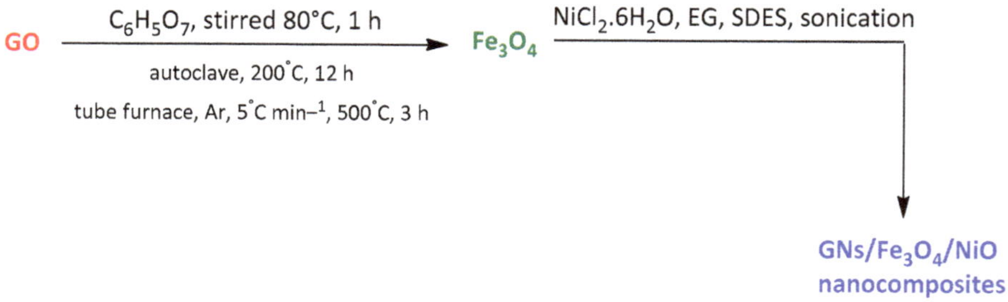

Scheme 33. Synthesis of GNs/Fe_2O_4/NiO nanocomposites.

The experimental findings indicated a notable improvement in both magnetic and adsorption performance for magnetic GNs/Fe_3O_4/NiO nanocomposites. The saturation magnetizations for the three types of nanocomposites, GNs/Fe_3O_4/NiO (S1), GNs/Fe_3O_4/NiO (S2) and GNs/Fe_3O_4/NiO (S3), were approximately 63.1, 43.3 and 22.4 emu g^{-1}, respectively.

The nanocomposites demonstrated elevated photocatalytic ability towards p–nitrophenol and RhB. Additionally, the degradation rate of S3 nanocomposites for p–nitrophenol and RhB were approximately 94.1% and 86.7%, respectively. Even after three uses, the degradation rate of S3 for p–nitrophenol and RhB remained above 90% and 84%, respectively. This suggests that the nanocomposites exhibit robust photocat–lytic performance and hold significant potential for future applications in the field of photocatalysis (Schemes 34 and 35).

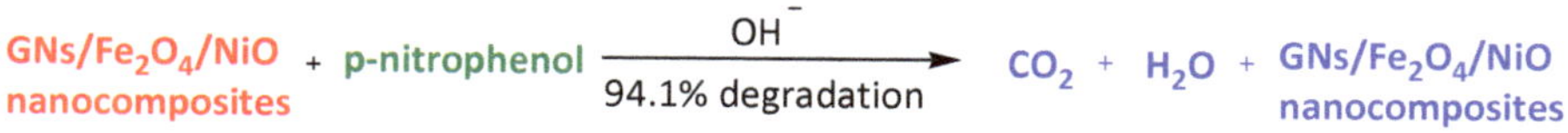

Scheme 34. Degradation of p–nitrophenol.

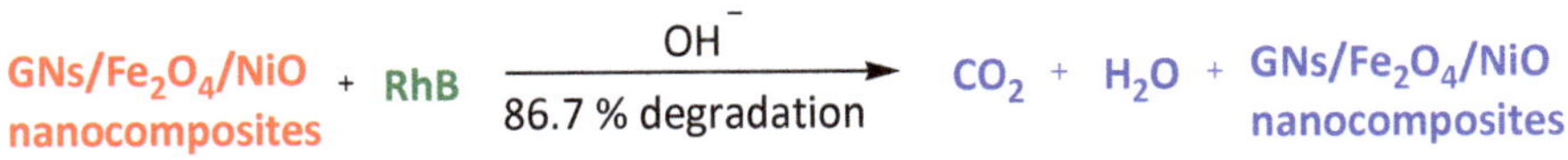

Scheme 35. Degradation of Rhodamine B.

Yang et al. synthesized super adsorbents by creating reduced graphene oxide (GO) on Fe oxide (GO/FeO•Fe_2O_3) for the adsorption of 1–naphthylamine, 1–naphthol and naphthalene with distinct polarities. GO was produced using a modified Hummers method from graphite flakes. Magnetic composites were formed through co-precipitation involving ferrous and ferric ions on the surfaces of GO or MWCNTs.

The adsorption followed the following order: 1–naphthylamine > 1–naphthol > naphthalene. The mechanism proposed was the electron donor–acceptor (EDA) interaction, with the adsorptive capability rising with an increase in dipole moment. In contrast to the combination of MWCNTs and iron oxide (MWCNTs/FeO•Fe_2O_3), we observed that the structure of the adsorbents significantly influenced the adsorption of these aromatic compounds [107] (Schemes 36–38).

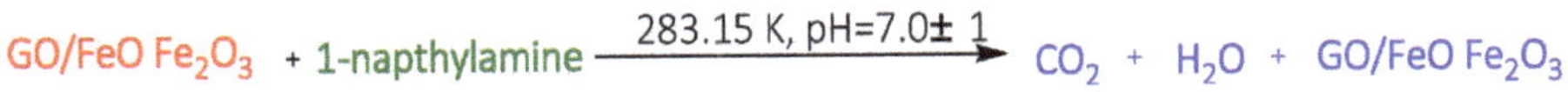

Scheme 36. Removal of 1-napthylamine.

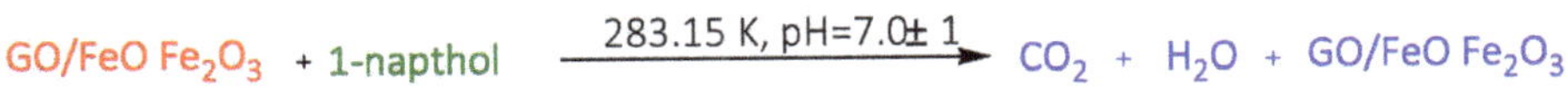

Scheme 37. Removal of 1-napthol.

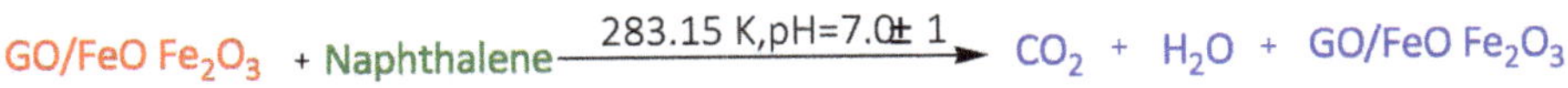

Scheme 38. Removal of naphthalene.

The synthesis procedure for the magnetic composite Fe_3O_4@graphene (FGC) involved the following steps: GO was dispersed in water through sonication to convert carboxylic acid groups into carboxylate anions. Subsequently, a solution of $FeCl_3$·and $FeCl_2$·was added to the GO suspension at 25 °C under N_2 with stirring. Following the ion exchange process, an ammonia solution was introduced dropwise to adjust the solution's pH to 10 for the synthesis of magnetite Fe_3O_4 nanoparticles [108] (Scheme 39).

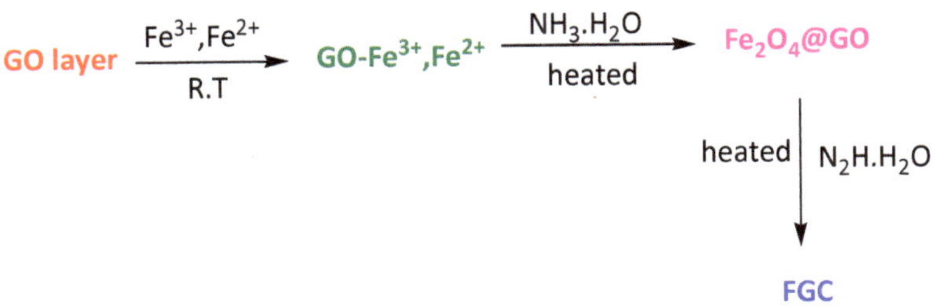

Scheme 39. Synthesis of FGC.

The adsorption of MB and Congo Red (CR) on the Fe_3O_4@graphene composite (FGC) was studied in batch. FGC demonstrated maximum adsorption capacities of 45.3 mg/g for MB) and 33.7 mg/g for CR. The sorption kinetics were represented by the second-order kinetic equation (Schemes 40 and 41).

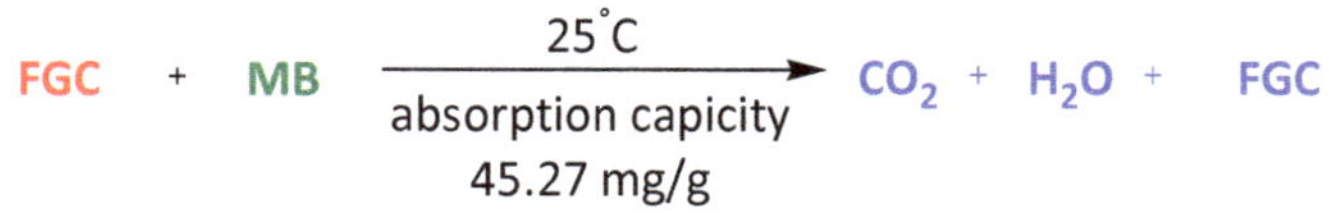

Scheme 40. Removal of methylene blue.

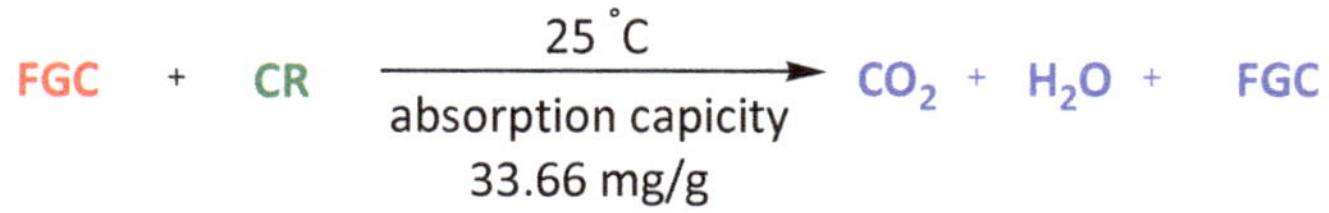

Scheme 41. Removal of Congo Red.

Li et al. obtained nanocomposites of magnetic $CoFe_2O_4$ functionalized graphene sheets ($CoFe_2O_4$–FGS) hydrothermally treating inorganic salts and thermally exfoliating graphene sheets. For studying the adsorption of $CoFe_2O_4$–FGS, the model compound chosen was the common contaminant methyl orange (MO). The $CoFe_2O_4$–FGS with magnetic separation capability showed significant adsorption ability (71.54 mg g^{-1}) for MO molecules, starting from 10 ppm. This suggests that $CoFe_2O_4$–FGS could be advantageous for applications related to separation and purification [109] (Scheme 42).

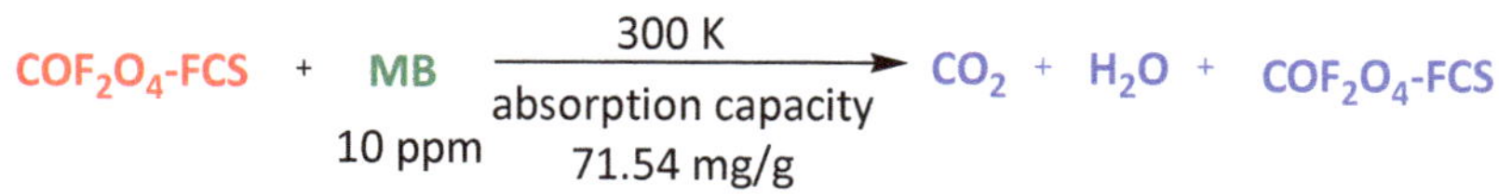

Scheme 42. Removal of methyl blue.

Wang et al. produced a graphene-based magnetic nanocomposite (G/Fe_3O_4) and utilized it to eliminate dye from water solutions. The synthesis of G/Fe_3O_4 involved the co-precipitation of Fe^{2+} and Fe^{3+} in alkaline conditions with graphene.

The adsorption process is rapid, achieving equilibrium within 30 min. The kinetic data were effectively modeled using a pseudo-second-order approach. Both Freundlich and Langmuir models were employed to analyze the isotherms. The collected data fitted a pseudo-second-order method. Even after undergoing five regenerations, the maximum amount of fuchsine that G/Fe_3O_4 can adsorb did not exhibit a notable decrease [110] (Scheme 43).

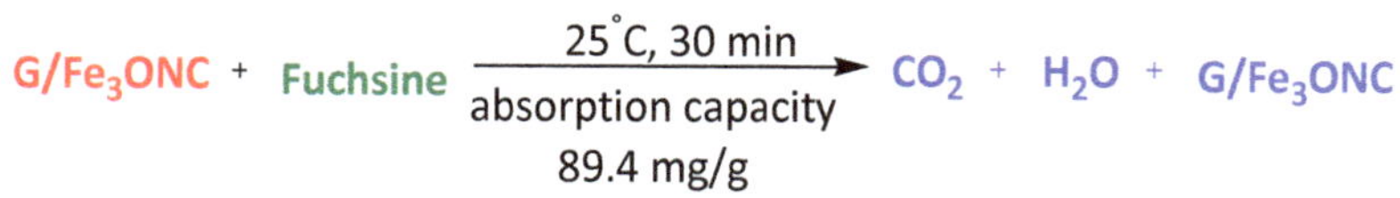

Scheme 43. Removal of fuchsine.

2.6. Magnetic Graphene Oxide Nanocomposites

Islam et al. introduced an innovative approach to creating a magnetic CNT-reduced GO (rGO) Ag nanocomposite. The Ag–rGO–PD–MCNT nanocomposite was created by reducing $AgNO_3$ in a rGO–PD–MCNT suspension using sodium citrate (Scheme 44).

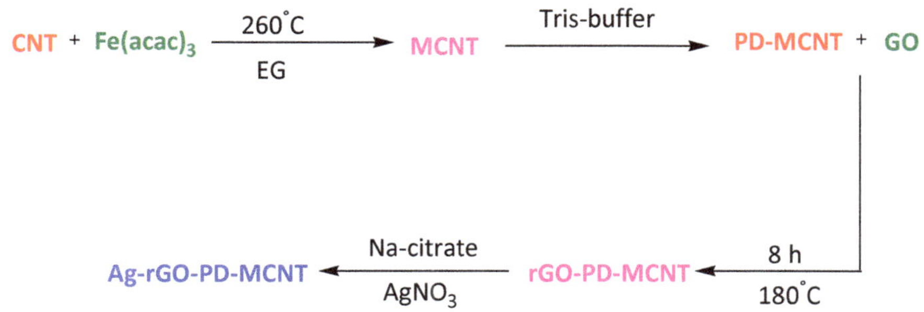

Scheme 44. Synthesis of Ag–rGO–PD–MCNT composite.

The synthesized nanocomposites exhibited significant catalytic efficacy alongside rapid and efficient adsorption, demonstrated across various pH levels during the elimination of a model 4-nitrophenol, MB and an aromatic nitro compound, toxic dye. Within 4 min, the absorbance at both wavelengths reached saturation and the distinctive yellow color of 4-nitrophenol almost completely disappeared, signifying an approximate 84% reduction. In the presence of the nanocomposites, the reduction of MB reached up to 96% within 10 min, resulting in the emergence of a colorless solution. The nanocomposites were magnetically isolatable, easily restored by desorption with water and ethanol, enabling their repeated use for over 15 cycles due to high recyclability [111] (Schemes 45 and 46).

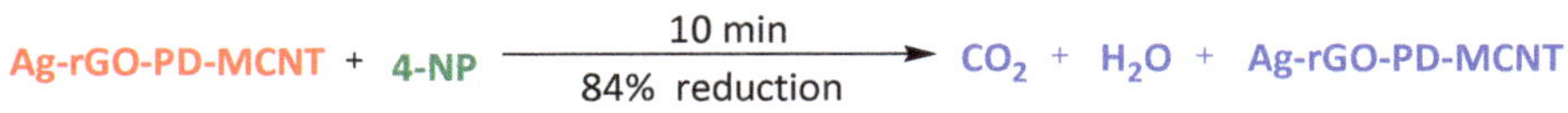

Scheme 45. Removal of 4-nitrophenol.

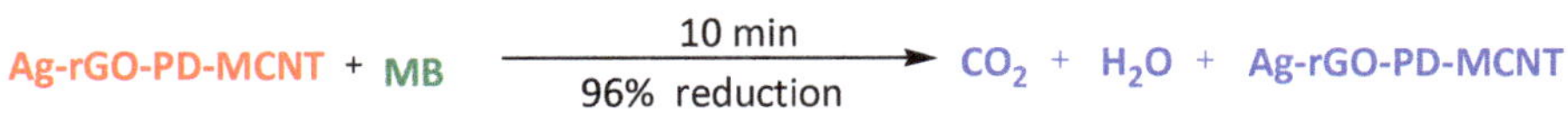

Scheme 46. Removal of methylene blue.

Chang et al. synthesized magnetic $CoFe_2O_4$/GO through a straightforward hydrothermal procedure. The structure, morphology and magnetic properties of the resulting materials were characterized using several techniques.

The $CoFe_2O_4$/GO composites were employed for removing RhB, MB and methyl orange (MO). Remarkably, there was minimal detection of adsorption for these three dyes on the initially synthesized $CoFe_2O_4$, validating that the adsorption capability of dyes on $CoFe_2O_4$/GO mainly arises from the presence and contribution of GO. A clear selective adsorption is observed, with the order being MO < RhB > MN. The Langmuir model reveals a remarkable adsorption of 355.9 mg/g for MB, 284.9 mg/g for RhB and 53.0 mg/g for MO. The presence of oxygenated groups in GO significantly influences its adsorption ability. The adsorption capacity is higher for carboxyl (–COOH) modifications on the GO surface compared to epoxy (–CH(O)CH–) or hydroxyl (–OH) species. This insight opens up a new avenue for designing adsorbent materials based on graphene oxide [112] (Schemes 47–49).

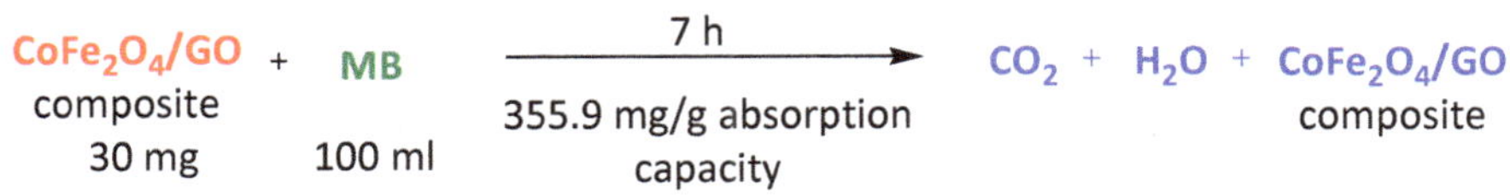

Scheme 47. Removal of methylene blue.

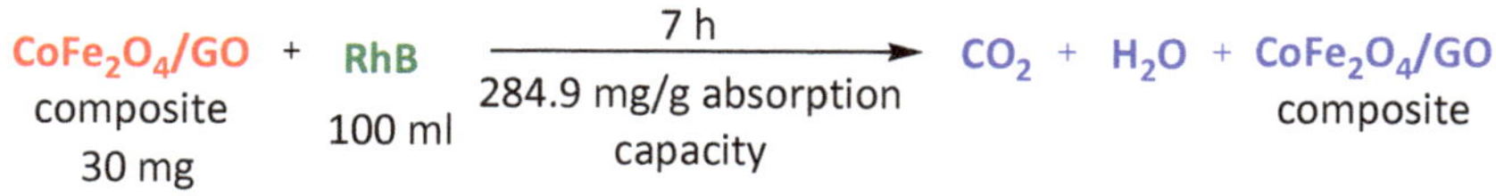

Scheme 48. Removal of Rhodanine B.

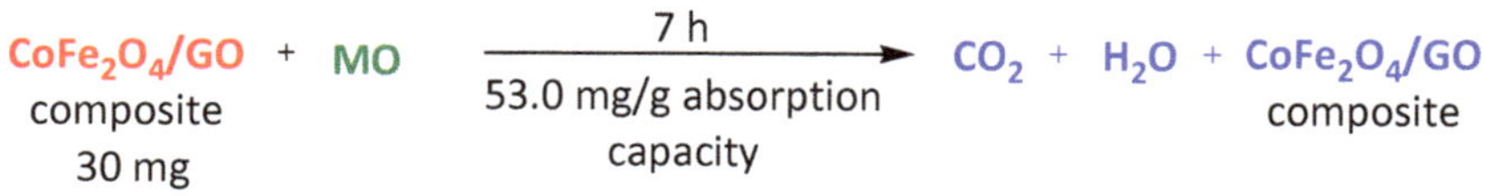

Scheme 49. Removal of methyl orange.

Chen et al. developed an innovative hybrid, Mn-doped Fe_3O_4 hollow microspheres on rGO (Mn-Fe_3O_4/rGO), using a simple solvothermal method. This was followed by reducing GO through $NaBH_4$ [113] (Scheme 50).

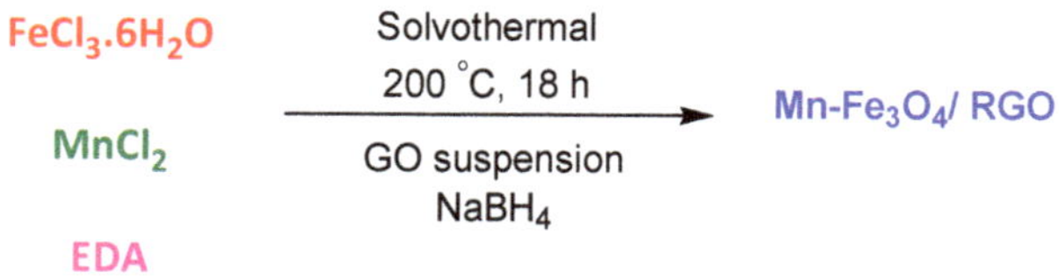

Scheme 50. Synthesis of Mn–Fe_3O_4/RGO.

The Mn–Fe_3O_4/RGO hybrid, produced in this manner, serves as a photo-Fenton material for RhB degradation, achieving a remarkable 96.4% efficiency with a minimal catalyst amount of 0.2 g/L. This happens when exposed to UV–visible light and with the presence of H_2O_2 at the ambient pH of around 6.5, all within an 80 min timeframe. Notably, removal efficiencies of 91% and 85% are observed at pH 11 and 2, respectively. Moreover, this innovative photo-Fenton material maintains a robust degradation of approximately 90% even after undergoing ten cycles (Scheme 51).

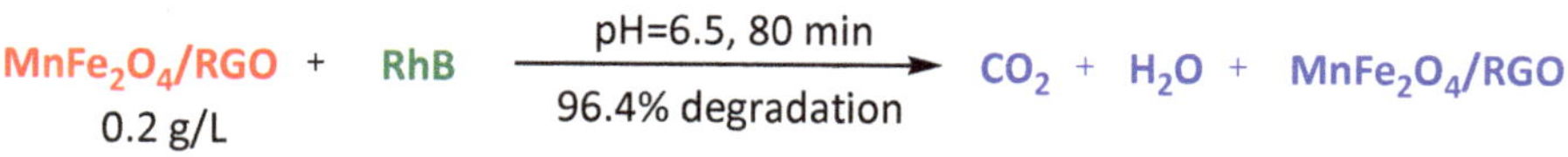

Scheme 51. Degradation of Rhodamine B.

Mishra and Mohanty synthesized GO sheets decorated with Fe_3O_4 nanoparticles for the purpose of removing organic pollutants. Fe_3O_4@GO nanocomposites were prepared by incorporating GO into double-distilled water through sonication. Subsequently, Fe_3O_4 powder was introduced into the solution, succeeded by ultrasonication. Ultimately, hydrazine hydrate was introduced into the blend, and then sonicated before undergoing centrifugation and subsequent washing [114] (Scheme 52).

Scheme 52. Synthesis of Fe_2O_4@GO.

The Fe_3O_4@GO nanocomposites exhibit superparamagnetic behavior, characterized by a magnetic saturation (MS) of 30.6 emu/g. Different concentrations of Fe_3O_4@GO nanocomposites were used to remove R6G from a water-based solution. Achieving a removal of 89%, these results hold significance for environmental considerations, demonstrating that Fe_3O_4@GO nanocomposites can effectively eliminate organic pollutants from water (Scheme 53).

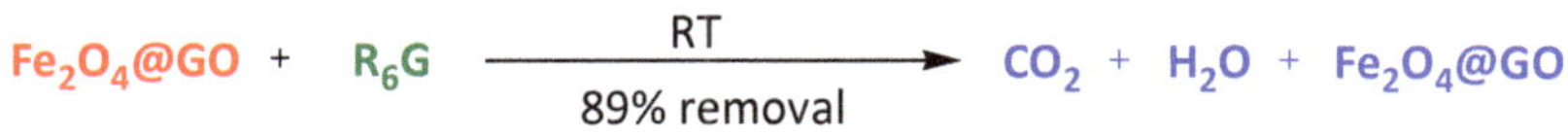

Scheme 53. Removal of Rhodamine 6G.

Bai et al. introduced a single-pot solvothermal technique for crafting hybrids that incorporate reduced ferrite (MFe_2O_4, M = Co, Mn, Ni, Zn) and rGO. The process involved utilizing metal and graphite oxide and ions (M^{2+} and Fe^{3+}) as initial components [115] (Scheme 54).

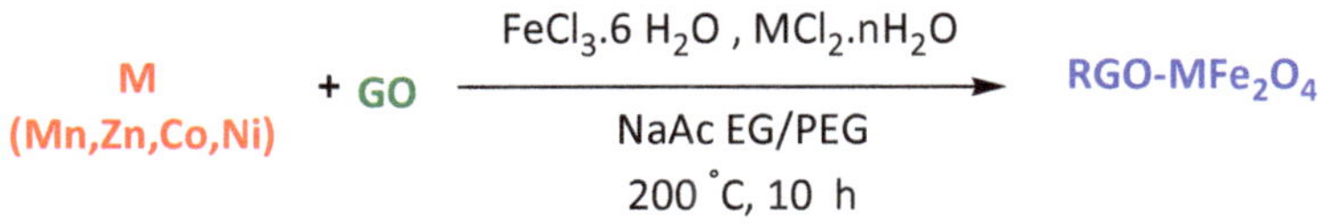

Scheme 54. Synthesis of RGO–MFe_2O_4.

The hybrids exhibit substantial saturation magnetization, along with reduced remanence and coercivity. Notably, these hybrids prove to be efficient adsorbents for eliminating dye pollutants. This shows that the hybrids can eliminate over 100% of MB and 92% of RhB with (5 mg/L) in 2 min, provided the hybrid concentration is 0.6 g/L (Schemes 55 and 56).

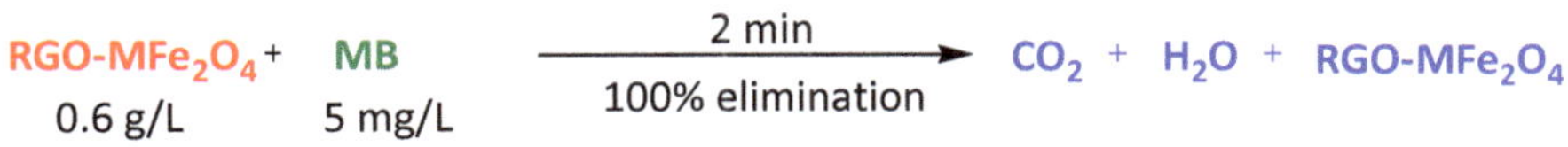

Scheme 55. Elimination of methylene blue.

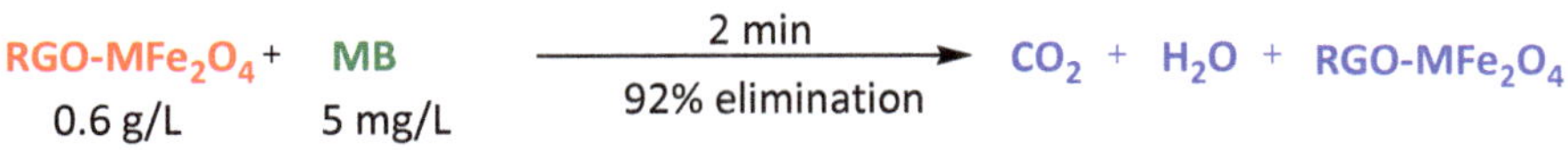

Scheme 56. Elimination of Rhodamine B.

Song et al. employed a single-step solvothermal technique to effectively create nanocomposites containing $CoFe_2O_4$ and reduced graphene oxide (CFG).

Due to favorable electrostatic interactions, CFG exhibits a significantly greater adsorption amount for anionic dyes in comparison to cationic dyes. The adsorption capacity for MO reaches a notable 263 mg g^{-1} at 25 °C, with the isotherms conforming to the Langmuir model. Additionally, CFG exhibits a specific saturation magnetization (Ms) of 32.8 emu g^{-1}, and the nanocomposites are quickly isolated using a magnetic field following the adsorption process. These results highlight the considerable promise of CFG in the treatment of real industrially used wastewater [116] (Scheme 57).

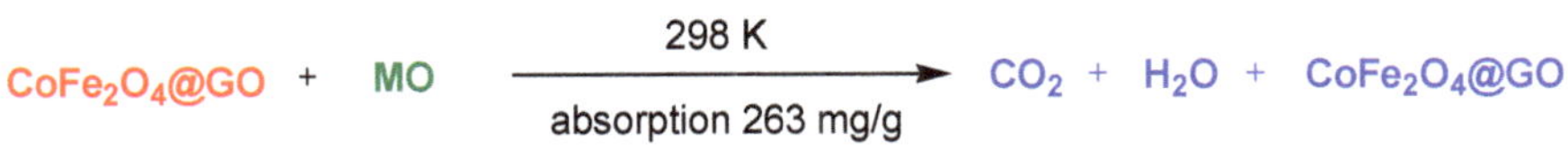

Scheme 57. Removal of methyl orange.

In summary, the efficiency of a variety of carbon-based nanoparticles and their composites towards the removal of various organic pollutants has been represented in Table 1.

Table 1. Efficiency of carbon-based nanoparticles/composites.

Type	Nanoparticles/Composites	Type of Organic Pollutants	Time	Absorption/ Elimination Efficiency of Nanomaterial	Ref.
Magnetic fullerene nanocomposites	Functionalized Magnetic Fullerene Nanocomposites	Methylene Blue Acid Blue 25	45 min 45 min	99.6% 97.01%	[88]
	Functionalized Magnetic Fullerene Nanocomposites	Ciprofloxacin	153 min	65 mg/L	[89]
Magnetic Carbon-dot Nanocomposites	Carbon-dot and magnetite-modified magnetic carbon nanotubes	Carbamazepine	3 h	65 mg/g	[90]
	Magnetic C-Dots	Methylene Blue	30 min	83%	[91]
Magnetic Carbon nanotube nanocomposites	Carbon nanotubes-incorporated MIL-88B-Fe	Phenol	30 min	55%	[93]
	Magnetic CNTs functionalized with polyethyleneimine	Alizarin Red S	40 min	94.6%	[95]
	Multiwalled carbon nanotubes	Rose Bengal	30 min	100%	[94]
	Magnetic Carbon-nanotube–Cyclodextrin composite	Methylene Blue	25 h	196.5 mg/g	[95]
	Magnetic graphitized MWCNTs modified with Chitosan	Crystal Violet	100 min	94.56–100%	[96]
	Magnetic polymers multiwall carbon nanotube nanocomposite	Orange (II) Sunset yellow FCF Amarnath	6 h 6 h 6 h	67.57 mg/g 85.47 mg/g 47.39 mg/g	[97]
	Magnetic-modified multiwalled carbon nanotubes	Methylene Blue Thioflavin Janus Green Crystal Violet	15 min 15 min 15 min 15 min	95% 95% 95% 100%	[98]
	Magnetic carbon nanotubes modified with guar gum GG-MWCNT-Fe_3O_4	Methylene Blue Neutral Red	120 min 20 min	37.4 mg/L 28.9 mg/L	[99]
	Magnetic multiwalled carbon nanotubes	Methylene Blue Neutral Red	60 min 60 min	42.3 mg/g 77.5 mg/g	[100]
Magnetic Graphite Nanocomposites	Magnetic graphite intercalation compounds as persulfate activators	Bisphenol A	75 min	99.3%	[101]
	Acid-resistant Magnetic Co/C nanocomposite	Rhodamine B Malachite Green	30 min 30 min	>99% >99%	[102]
	Magnetic Fe_2O_4–graphite composite	Levofloxacin	15 min	80%	[103]
Magnetic Graphene Nanocomposites	Fe_2O_4/porous graphene nanocomposite	Methyl Violet	5 min	460 mg/g	[104]
	Magnetic sponge of graphene (Fe_3O_4-GS)	Methyl blue	5 h	526 mg/g	[105]
	Magnetic photocatalyst of graphene with Fe_3O_4 and NiO GNs/Fe_3O_4/NiO nanocomposite	p-Nitrophenol Rhodamine B	20 min	94.1% 86.7%	[106]
	Reduced graphene Oxide (GO) on Fe oxide (GO/FeO.Fe_2O_3)	1-Naphthylamine 1-Napthol Napthalene		1.45 mmol/g 1.13 mmol/g 1.05 mmol/g	[107]
	Magnetic composite Fe_3O_4@graphene	Methylene Blue Congo Red	30 min 30 min	45.27 mg/g 33.66 mg/g	[108]
	Nanocomposites of magnetic $CoFe_2O_4$ functionalized graphene sheets ($CoFe_2O_4$-FGS)	Methyl Orange	60 min	71.54 mg/g	[109]
	Graphene-based magnetic nanocomposite (G/Fe_3O_4)	Fuchsine	30 min	89.4 mg/g	[110]
Magnetic graphene oxide nanocomposites	Magnetic RCNT-CD	4-Nitrophenol Methylene Blue	10 min 10 min	84% 96%	[111]
	Magnetic $CoFe_2O_4$/GO	Methylene Blue Rhodamine B Methyl Orange	7 h 7 h 7 h	355.9 mg/g 284.9 mg/g 53.0 mg/g	[112]
	Mn-doped Fe_3O_4 hollow microspheres on rGO	Rhodamine B	80 min	96.4%	[113]
	GO sheets with Fe_3O_4 nanoparticles	Rhodamine 6G	5 min	89%	[114]
	Reduced graphene oxide (RGO)-supported ferrite (MFe_2O_4, M = Mn, Zn, Co, Ni)	Methylene Blue Rhodamine B	2 min 2 min	100% 92%	[115]
	Nanocomposite containing $CoFe_2O_4$ and reduced graphene oxide	Methyl Orange	2 h	263 mg/g	[116]

3. Conclusions and Perspectives

A comprehensive examination of carbon nano-adsorbents, including graphene, graphene oxide, fullerene, carbon dots, graphite and carbon nanotubes has demonstrated their remarkable ability to eliminate organic pollutants, making them an excellent choice for environmental clean-up. The effectiveness of removing organic pollutants using these materials is attributed to their recyclable nature, low cytotoxicity, homogeneous nanoparticle deposition, and environmentally friendly procedures. While these materials have found extensive use in practical applications for eliminating organic contaminants from wastewater, the cost of carbon nanocomposites and the uncertainty regarding the potential hazards of nanosized carbon compounds persist in environmental systems. Despite the significant impact that these materials can have, there is a need for the development of more green synthesis methods and economical techniques for characterizing them. In the current scenario, the scientific community should shift its focus towards the creation of effective methods utilizing biological agents such as microbes and enzymes. These agents have the potential to break down a variety of organic pollutants and could offer more sustainable solutions.

Author Contributions: B.A.: conceptualization, writing—original draft; writing—reviewing and editing; R.R.: supervision; V.: writing—original draft; M.A.: writing—reviewing and editing; S.A.C.C.: supervision; writing—reviewing and editing. All authors have read and agreed to the published version of the manuscript.

Funding: SACC also acknowledges (FCT/MCTES, Fundação para a Ciência e Tecnologia and Ministério da Ciência, Tecnologia e Ensino Superior) for projects UIDB/50006/2020 and UIDP/50006/2020FCT and for the Scientific Employment Stimulus—Institutional Call (https://doi.org/10.54499/CEECINST/00102/2018/CP1567/CT0026).

Institutional Review Board Statement: Not applicable.

Informed Consent Statement: Not applicable.

Data Availability Statement: The data used in this paper are the articles from other authors that were consulted and are available online.

Acknowledgments: The authors would like to express their sincere thanks to the management of Presidency University, Rajanukunte, Itgalpura, Bangalore, India, Baba Masthnath University, Rohtak, Haryana, India, Universidade NOVA de Lisboa, and Islamic Azad University, Shoushtar, Iran for providing the required facilities to write and submit the article for publication. SACC also acknowledges (FCT/MCTES, Fundação para a Ciência e Tecnologia and Ministério da Ciência, Tecnologia e Ensino Superior) for projects UIDB/50006/2020 and UIDP/50006/2020FCT and for the Scientific Employment Stimulus—Institutional Call (DOI 10.54499/CEECINST/00102/2018/CP1567/CT0026).

Conflicts of Interest: The authors declare no conflicts of interest.

References

1. Narain-Ford, D.M.; Bartholomeus, R.P.; Dekker, S.C.; van Wezel, A.P. Natural Purification Through Soils: Risks and Opportunities of Sewage Effluent Reuse in Sub-surface Irrigation. In *Reviews of Environmental Contamination and Toxicology*; Springer: Berlin/Heidelberg, Germany, 2020; Volume 250, pp. 85–117.
2. Pham-Duc, P.; Nguyen-Viet, H.; Hattendorf, J.; Cam, P.D.; Zurbrügg, C.; Zinsstag, J.; Odermatt, P. Diarrhoeal diseases among adult population in an agricultural community Hanam province, Vietnam, with high wastewater and excreta re-use. *BMC Public Health* **2014**, *14*, 1–14. [CrossRef]
3. Yang, J.; Jia, R.S.; Gao, Y.L.; Wang, W.F.; Cao, P.Q. The reliability evaluation of reclaimed water reused in power plant project. *IOP Conf. Ser. Earth Environ. Sci.* **2017**, *100*, 012189. [CrossRef]
4. Clemmens, A.J.; Allen, R.G.; Burt, C.M. Technical concepts related to conservation of irrigation and rainwater in agricultural systems. *Water Resour. Res.* **2008**, *44*, W00E03. [CrossRef]
5. Sato, T.; Qadir, M.; Yamamoto, S.; Endo, T.; Zahoor, A. Global, regional, and country level need for data on wastewater generation, treatment, and use. *Agric. Water Manag.* **2013**, *130*, 1–13. [CrossRef]
6. Dickin, S.K.; Schuster-Wallace, C.J.; Qadir, M.; Pizzacalla, K. A review of health risks and pathways for exposure to wastewater Use in Agriculture. *Environ. Health Perspect.* **2016**, *124*, 900–909. [CrossRef]
7. Duan, B.; Zhang, W.; Zheng, H.; Wu, C.; Zhang, Q.; Bu, Y. Comparison of health risk assessments of heavy metals and as in sewage sludge from wastewater treatment plants (WWTPs) for adults and children in the urban district of Taiyuan, China. *Int. J. Environ. Res. Public Health* **2017**, *14*, 1194. [CrossRef]
8. Carr, R. Who guidelines for safe wastewater use—More than just numbers. *Irrig. Drain.* **2005**, *54* (Suppl. S1), S103–S111. [CrossRef]

9. Pedrero, F.; Kalavrouziotis, I.; Alarcón, J.J.; Koukoulakis, P.; Asano, T. Use of treated municipal wastewater in irrigated agriculture—Review of some practices in Spain and Greece. *Agric. Water Manag.* **2010**, *97*, 1233–1241. [CrossRef]
10. Katsoyiannis, I.A.; Gkotsis, P.; Castellana, M.; Cartechini, F.; Zouboulis, A.I. Production of demineralized water for use in thermal power stations by advanced treatment of secondary wastewater effluent. *J. Environ. Manag.* **2017**, *190*, 132–139. [CrossRef]
11. Chaoua, S.; Boussaa, S.; El Gharmali, A.; Boumezzough, A. Impact of irrigation with wastewater on accumulation of heavy metal in soil and crops in the region of Marrakech in Morocco. *J. Saudi Soc. Agric. Sci.* **2019**, *18*, 429–436. [CrossRef]
12. Poustie, A.; Yang, Y.; Verburg, P.; Pagilla, K.; Hanigan, D. Reclaimed wastewater as a viable water source for agricultural irrigation: A review of food crop growth inhibition and promotion in the context of environmental change. *Sci. Total Environ.* **2020**, *739*, 139756. [CrossRef] [PubMed]
13. Balkhair, K.S. Microbial contamination of vegetable crop and soil profile in arid regions under controlled application of domestic wastewater. *Saudi J. Biol. Sci.* **2016**, *23*, S83–S92. [CrossRef]
14. Contreras, J.D.; Meza, R.; Siebe, C.; Rodríguez-Dozal, S.; López-Vidal, Y.A.; Castillo-Rojas, G.; Amieva, R.I.; Solano-Gálvez, S.G.; Mazari-Hiriart, M.; Silva-Magaña, M.A.; et al. Health risks from exposure to untreated wastewater used for irrigation in the Mezquital Valley, Mexico: A 25-year update. *Water Res.* **2017**, *123*, 834–850. [CrossRef] [PubMed]
15. Singh, A.; Sawant, M.; Kamble, S.J.; Herlekar, M.; Starkl, M.; Aymerich, E.; Kazmi, A. Performance evaluation of a decentralized wastewater treatment system in India. *Environ. Sci. Pollut. Res.* **2019**, *26*, 21172–21188. [CrossRef]
16. Ungureanu, N.; Vladut, V.; Dincă, M.; Zăbavă, B.-Ș.; Vlăduţ, V. Reuse of wastewater for irrigation. A sustainable practice in arid and semi-arid regions. In Proceedings of the 7th International Conference on Thermal Equipment, Renewable Energy and Rural Development (TE-RE-RD), Drobeta-Turnu Severin, Romania, 31 May–2 June 2018.
17. Angelakis, A.N.; Snyder, S.A. Wastewater treatment and reuse: Past, present, and future. *Water* **2015**, *7*, 4887–4895. [CrossRef]
18. Jeong, H.; Kim, H.; Jang, T. Irrigation water quality standards for indirect wastewater reuse in agriculture: A contribution toward sustainable wastewater reuse in South korea. *Water* **2016**, *8*, 169. [CrossRef]
19. Kimura, K.; Honoki, D.; Sato, T. Effective physical cleaning and adequate membrane flux for direct membrane filtration (DMF) of municipal wastewater: Up-concentration of organic matter for efficient energy recovery. *Sep. Purif. Technol.* **2017**, *181*, 37–43. [CrossRef]
20. Ding, W.; Cheng, S.; Yu, L.; Huang, H. Effective swine wastewater treatment by combining microbial fuel cells with flocculation. *Chemosphere* **2017**, *182*, 567–573. [CrossRef]
21. Rott, E.; Minke, R.; Steinmetz, H. Removal of phosphorus from phosphonate-loaded industrial wastewaters via precipitation/flocculation. *J. Water Process Eng.* **2017**, *17*, 188–196. [CrossRef]
22. Lin, C.Y.; Chiang, C.C.; Thi Nguyen, M.L.; Lay, C.H. Enhancement of fermentative biohydrogen production from textile desizing wastewater via coagulation-pretreatment. *Int. J. Hydrogen Energy* **2017**, *42*, 12153–12158. [CrossRef]
23. Cristóvão, R.O.; Gonçalves, C.; Botelho, C.M.; Martins, R.J.E.; Boaventura, R.A.R. Chemical oxidation of fish canning wastewater by Fenton's reagent. *J. Environ. Chem. Eng.* **2014**, *2*, 2372–2376. [CrossRef]
24. Li, R.; Yang, C.; Chen, H.; Zeng, G.; Yu, G.; Guo, J. Removal of triazophos pesticide from wastewater with Fenton reagent. *J. Hazard. Mater.* **2009**, *167*, 1028–1032. [CrossRef] [PubMed]
25. Defaei, M.; Taheri-Kafrani, A.; Miroliaei, M.; Yaghmaei, P. Improvement of stability and reusability of α-amylase immobilized on naringin functionalized magnetic nanoparticles: A robust nanobiocatalyst. *Int. J. Biol. Macromol.* **2018**, *113*, 354–360. [CrossRef]
26. Berberidou, C.; Kitsiou, V.; Lambropoulou, D.A.; Antoniadis, A.; Ntonou, E.; Zalidis, G.C.; Poulios, I. Evaluation of an alternative method for wastewater treatment containing pesticides using solar photocatalytic oxidation and constructed wetlands. *J. Environ Manag.* **2017**, *195*, 133–139. [CrossRef]
27. Patel, H.; Vashi, R.T. Batch Adsorption Treatment of Textile Wastewater. In *Characterization and Treatment of Textile Wastewater*; Elsevier: Amsterdam, The Netherlands, 2015; pp. 111–125. [CrossRef]
28. Kyzas, G.Z.; Fu, J.; Lazaridis, N.K.; Bikiaris, D.N.; Matis, K.A. New approaches on the removal of pharmaceuticals from wastewaters with adsorbent materials. *J. Mol. Liq.* **2015**, *209*, 87–93. [CrossRef]
29. de Caprariis, B.; De Filippis, P.; Hernandez, A.D.; Petrucci, E.; Petrullo, A.; Scarsella, M.; Turchi, M. Pyrolysis wastewater treatment by adsorption on biochars produced by poplar biomass. *J. Environ. Manag.* **2017**, *197*, 231–238. [CrossRef] [PubMed]
30. Suárez-Iglesias, O.; Collado, S.; Oulego, P.; Díaz, M. Graphene-family nanomaterials in wastewater treatment plants. *Chem. Eng. J.* **2017**, *313*, 121–135. [CrossRef]
31. Delkash, M.; Ebrazi Bakhshayesh, B.; Kazemian, H. Using zeolitic adsorbents to cleanup special wastewater streams: A review. *Microporous Mesoporous Mater.* **2015**, *214*, 224–241. [CrossRef]
32. Wang, S.; Peng, Y. Natural zeolites as effective adsorbents in water and wastewater treatment. *Chem. Eng. J.* **2010**, *156*, 11–24. [CrossRef]
33. Umejuru, E.C.; Mashifana, T.; Kandjou, V.; Amani-Beni, M.; Sadeghifar, H.; Fayazi, M.; Karimi-Maleh, H.; Sithole, N.T. Application of zeolite based nanocomposites for wastewater remediation: Evaluating newer and environmentally benign approaches. *Environ. Res.* **2023**, *231*, 116073. [CrossRef]
34. Bialczyk, J.; Natkański, P.; Kuśtrowski, P.; Czaja-Prokop, U.; Bober, B.; Kaminski, A. Removal of cyanobacterial anatoxin-a from water by natural clay adsorbents. *Appl. Clay Sci.* **2017**, *148*, 17–24. [CrossRef]
35. Khan, T.A.; Dahiya, S.; Ali, I. Use of kaolinite as adsorbent: Equilibrium, dynamics and thermodynamic studies on the adsorption of Rhodamine B from aqueous solution. *Appl. Clay Sci.* **2012**, *69*, 58–66. [CrossRef]

36. Magriotis, Z.M.; Leal, P.V.B.; de Sales, P.F.; Papini, R.M.; Viana, P.R.M.; Arroyo, P.A. A comparative study for the removal of mining wastewater by kaolinite, activated carbon and beta zeolite. *Appl. Clay Sci.* **2014**, *91–92*, 55–62. [CrossRef]
37. Struijk, M.; Rocha, F.; Detellier, C. Novel thio-kaolinite nanohybrid materials and their application as heavy metal adsorbents in wastewater. *Appl. Clay Sci.* **2017**, *150*, 192–201. [CrossRef]
38. Gil, A.; Assis, F.C.C.; Albeniz, S.; Korili, S.A. Removal of dyes from wastewaters by adsorption on pillared clays. *Chem. Eng. J.* **2011**, *168*, 1032–1040. [CrossRef]
39. Georgescu, A.M.; Nardou, F.; Zichil, V.; Nistor, I.D. Adsorption of lead(II) ions from aqueous solutions onto Cr-pillared clays. *Appl. Clay Sci.* **2018**, *152*, 44–50. [CrossRef]
40. Rodrigues Mota, T.L.; Marques de Oliveira, A.P.; Nunes, E.H.M.; Houmard, M. Simple process for preparing mesoporous sol-gel silica adsorbents with high water adsorption capacities. *Microporous Mesoporous Mater.* **2017**, *253*, 177–182. [CrossRef]
41. Banaei, A.; Samadi, S.; Karimi, S.; Vojoudi, H.; Pourbasheer, E.; Badiei, A. Synthesis of silica gel modified with 2,2′-(hexane-1,6-diylbis(oxy)) dibenzaldehyde as a new adsorbent for the removal of Reactive Yellow 84 and Reactive Blue 19 dyes from aqueous solutions: Equilibrium and thermodynamic studies. *Powder Technol.* **2017**, *319*, 60–70. [CrossRef]
42. Asfaram, A.; Ghaedi, M.; Ahmadi Azqhandi, M.H.; Goudarzi, A.; Hajati, S. Ultrasound-assisted binary adsorption of dyes onto Mn@ CuS/ZnS-NC-AC as a novel adsorbent: Application of chemometrics for optimization and modeling. *J. Ind. Eng. Chem.* **2017**, *54*, 377–388. [CrossRef]
43. Mazaheri, H.; Ghaedi, M.; Asfaram, A.; Hajati, S. Performance of CuS nanoparticle loaded on activated carbon in the adsorption of methylene blue and bromophenol blue dyes in binary aqueous solutions: Using ultrasound power and optimization by central composite design. *J. Mol. Liq.* **2016**, *219*, 667–676. [CrossRef]
44. Choleva, T.G.; Gatselou, V.A.; Tsogas, G.Z.; Giokas, D.L. Intrinsic peroxidase-like activity of rhodium nanoparticles, and their application to the colorimetric determination of hydrogen peroxide and glucose. *Microchim. Acta* **2018**, *185*, 22. [CrossRef]
45. Novais, R.M.; Ascensão, G.; Tobaldi, D.M.; Seabra, M.P.; Labrincha, J.A. Biomass fly ash geopolymer monoliths for effective methylene blue removal from wastewaters. *J. Clean. Prod.* **2018**, *171*, 783–794. [CrossRef]
46. Habibzadeh, M.; Chaibakhsh, N.; Naeemi, A.S. Optimized treatment of wastewater containing cytotoxic drugs by living and dead biomass of the freshwater microalga, *Chlorella vulgaris*. *Ecol. Eng.* **2018**, *111*, 85–93. [CrossRef]
47. Vukelic, D.; Boskovic, N.; Agarski, B.; Radonic, J.; Budak, I.; Pap, S.; Sekulic, M.T. Eco-design of a low-cost adsorbent produced from waste cherry kernels. *J. Clean. Prod.* **2018**, *174*, 1620–1628. [CrossRef]
48. Anjum, M.; Miandad, R.; Waqas, M.; Gehany, F.; Barakat, M.A. Remediation of wastewater using various nano-materials. *Arab. J. Chem.* **2019**, *12*, 4897–4919. [CrossRef]
49. Tripathi, A.; Rawat Ranjan, M. Heavy Metal Removal from Wastewater Using Low Cost Adsorbents. *J. Bioremediat. Biodegrad.* **2015**, *6*, 315. [CrossRef]
50. Abbaszadeh, S.; Wan Alwi, S.R.; Webb, C.; Ghasemi, N.; Muhamad, I.I. Treatment of lead-contaminated water using activated carbon adsorbent from locally available papaya peel biowaste. *J. Clean. Prod.* **2016**, *118*, 210–222. [CrossRef]
51. Reck, I.M.; Paixão, R.M.; Bergamasco, R.; Vieira, M.F.; Vieira, A.M.S. Removal of tartrazine from aqueous solutions using adsorbents based on activated carbon and *Moringa oleifera* seeds. *J. Clean. Prod.* **2018**, *171*, 85–97. [CrossRef]
52. Lladó, J.; Gil, R.R.; Lao-Luque, C.; Solé-Sardans, M.; Fuente, E.; Ruiz, B. Highly microporous activated carbons derived from biocollagenic wastes of the leather industry as adsorbents of aromatic organic pollutants in water. *J. Environ. Chem. Eng.* **2017**, *5*, 2090–2100. [CrossRef]
53. Bendi, A.; Dharma Rao, G.B.; Sharma, N.; Singh, M.P. Results in Chemistry $CoFe_2O_4/Cu(OH)_2$ Nanocomposite: Expeditious and magnetically recoverable heterogeneous catalyst for the four component Biginelli/transesterification reaction and their DFT studies. *Results Chem.* **2021**, *3*, 100202. [CrossRef]
54. Bendi, A.; Dharma Rao, G.B.; Nancy; Nagakalyan, S. Synthesis and DFT studies of 1,2-disubstituted benzimidazoles using expeditious and magnetically recoverable $CoFe_2O_4/Cu(OH)_2$ nanocomposite under solvent-free condition. *J. Saudi Chem. Soc.* **2021**, *25*, 101394. [CrossRef]
55. Farghali, A.A.; Bahgat, M.; Enaiet Allah, A.; Khedr, M.H. Adsorption of Pb(II) ions from aqueous solutions using copper oxide nanostructures. *Beni-Suef Univ. J. Basic Appl. Sci.* **2013**, *2*, 61–71. [CrossRef]
56. Dissanayake, M.A.K.L.; Divarathna, H.K.D.W.M.N.; Dissanayake, C.B.; Senadeera, G.K.R.; Ekanayake, P.M.P.C.; Thotawattage, C.A. An innovative TiO_2 nanoparticle/nanofibre/nanoparticle, three layer composite photoanode for efficiency enhancement in dye-sensitized solar cells. *J. Photochem. Photobiol. A Chem.* **2016**, *322–323*, 110–118. [CrossRef]
57. Muhamad, S.U.; Idris, N.H.; Yusoff, H.M.; Din, M.F.M.; Majid, S.R. In-situ encapsulation of nickel nanoparticles in polypyrrole nanofibres with enhanced performance for supercapacitor. *Electrochim. Acta* **2017**, *249*, 9–15. [CrossRef]
58. Carabineiro, S.A.C.; Dharma Rao, G.B.; Singh, L.; Anjaneyulu, B.; Afshari, M. $CuFe_2O_4$ Magnetic Nanoparticles as Heterogeneous Catalysts for Synthesis of Dihydropyrimidinones as Inhibitors of SARS-CoV-2 Surface Proteins—Insights from Molecular Docking Studies. *Processes* **2023**, *11*, 2294. [CrossRef]
59. Asfaram, A.; Ghaedi, M.; Goudarzi, A.; Rajabi, M. Response surface methodology approach for optimization of simultaneous dye and metal ion ultrasound-assisted adsorption onto Mn doped Fe_3O_4-NPs loaded on AC: Kinetic and isothermal studies. *Dalton Trans.* **2015**, *44*, 14707–14723. [CrossRef]
60. Ganesan, V.; Louis, C.; Damodaran, S.P. Graphene oxide-wrapped magnetite nanoclusters: A recyclable functional hybrid for fast and highly efficient removal of organic dyes from wastewater. *J. Environ. Chem. Eng.* **2018**, *6*, 2176–2190. [CrossRef]

61. Asfaram, A.; Ghaedi, M.; Hajati, S.; Goudarzi, A. Ternary dye adsorption onto MnO_2 nanoparticle-loaded activated carbon: Derivative spectrophotometry and modeling. *RSC Adv.* **2015**, *5*, 72300–72320. [CrossRef]
62. Anjaneyulu, B.; Chinmay; Chauhan, V.; Sonia, A.C.C.; Mozhgan, A. Recent advances on zinc ferrite and its derivatives as the forerunner of the nanomaterials in catalytic applications. *J. Inorg. Organomet. Polym.* **2023**, 1–21. [CrossRef]
63. Siqueira, J.R.; Oliveira, O.N. 9–Carbon-Based Nanomaterials. In *Nanostructures*; Elsevier: Oxford, UK, 2017. [CrossRef]
64. Ren, X.; Zeng, G.; Tang, L.; Wang, J.; Wan, J.; Feng, H.; Song, B.; Huang, C.; Tang, X. Effect of exogenous carbonaceous materials on the bioavailability of organic pollutants and their ecological risks. *Soil Biol. Biochem.* **2018**, *116*, 70–81. [CrossRef]
65. Yang, Y.; Yang, X.; Yang, Y.; Yuan, Q. Aptamer-functionalized carbon nanomaterials electrochemical sensors for detecting cancer relevant biomolecules. *Carbon* **2018**, *129*, 380–395. [CrossRef]
66. Malhotra, B.D.; Ali, M.A. Functionalized Carbon Nanomaterials for Biosensors. In *Nanomaterials for Biosensors*; Elsevier: Amsterdam, The Netherlands, 2018; Chapter 2; pp. 75–103. [CrossRef]
67. Kwon, Y.J.; Kim, Y.; Jeon, H.; Cho, S.; Lee, W.; Lee, J.U. Graphene/carbon nanotube hybrid as a multi-functional interfacial reinforcement for carbon fiber-reinforced composites. *Compos. Part B Eng.* **2017**, *122*, 23–30. [CrossRef]
68. Imani Yengejeh, S.; Kazemi, S.A.; Öchsner, A. Carbon nanotubes as reinforcement in composites: A review of the analytical, numerical and experimental approaches. *Comput. Mater. Sci.* **2017**, *136*, 85–101. [CrossRef]
69. Chen, Y.F.; Tan, Y.J.; Li, J.; Hao, Y.B.; Shi, Y.D.; Wang, M. Graphene oxide-assisted dispersion of multi-walled carbon nanotubes in biodegradable Poly(ε-caprolactone) for mechanical and electrically conductive enhancement. *Polym. Test.* **2018**, *65*, 387–397. [CrossRef]
70. Deb, A.; Vimala, R. Camptothecin loaded graphene oxide nanoparticle functionalized with polyethylene glycol and folic acid for anticancer drug delivery. *J. Drug Deliv. Sci. Technol.* **2018**, *43*, 333–342. [CrossRef]
71. Ahmed, W.; Elhissi, A.; Dhanak, V.; Subramani, K. Carbon nanotubes: Applications in cancer therapy and drug delivery research. In *Emerging Nanotechnologies in Dentistry*; Elsevier: Amsterdam, The Netherlands, 2017; pp. 371–389. [CrossRef]
72. Ioniță, M.; Crică, L.E.; Voicu, S.I.; Dinescu, S.; Miculescu, F.; Costache, M.; Iovu, H. Synergistic effect of carbon nanotubes and graphene for high performance cellulose acetate membranes in biomedical applications. *Carbohydr. Polym.* **2018**, *183*, 50–61. [CrossRef]
73. Li, Z.-F.; Xin, L.; Yang, F.; Liu, Y.; Liu, Y.; Zhang, H.; Stanciu, L.; Xie, J. Hierarchical polybenzimidazole-grafted graphene hybrids as supports for Pt nanoparticle catalysts with excellent PEMFC performance. *Nano Energy* **2015**, *16*, 281–292. [CrossRef]
74. Jeng, K.T.; Hsu, N.Y.; Chien, C.C. Synthesis and evaluation of carbon nanotube-supported RuSe catalyst for direct methanol fuel cell cathode. *Int. J. Hydrogen Energy* **2011**, *36*, 3997–4006. [CrossRef]
75. Teow, Y.H.; Mohammad, A.W. New generation nanomaterials for water desalination: A review. *Desalination* **2019**, *451*, 2–17. [CrossRef]
76. Thines, R.K.; Mubarak, N.M.; Nizamuddin, S.; Sahu, J.N.; Abdullah, E.C.; Ganesan, P. Application potential of carbon nanomaterials in water and wastewater treatment: A review. *J. Taiwan Inst. Chem. Eng.* **2017**, *72*, 116–133. [CrossRef]
77. Ray, S.C.; Jana, N.R. Application of Carbon-Based Nanomaterials for Removal of Biologically Toxic Materials. In *Carbon Nanomaterials for Biological and Medical Applications*; Elsevier: Amsterdam, The Netherlands, 2017; pp. 43–86. [CrossRef]
78. Kong, L.P.; Gan, X.J.; bin Ahmad, A.L.; Hamed, B.H.; Evarts, E.R.; Ooi, B.S.; Lim, J. Design and synthesis of magnetic nanoparticles augmented microcapsule with catalytic and magnetic bifunctionalities for dye removal. *Chem. Eng. J.* **2012**, *197*, 350–358. [CrossRef]
79. Lotfi Zadeh Zhad, H.R.; Aboufazeli, F.; Sadeghi, O.; Amani, V.; Najafi, E.; Tavassoli, N. Tris(2-Aminoethyl)amine-functionalized Femagnetic nanoparticles as a selective sorbent for separation of silver and gold ions in different pHs. *J. Chem.* **2013**, *2013*, 482793. [CrossRef]
80. Carreño, N.L.V.; Escote, M.T.; Valentini, A.; McCafferty, L.; Stolojan, V.; Beliatis, M.; Mills, C.A.; Rhodes, R.; Smith, C.T.G.; Silva, S.R.P. Adsorbent 2D and 3D carbon matrices with protected magnetic iron nanoparticles. *Nanoscale* **2015**, *7*, 17441–17449. [CrossRef]
81. Chen, C.L.; Wang, X.K.; Nagatsu, M. Europium adsorption on multiwall carbon nanotube/iron oxide magnetic composite in the presence of polyacrylic acid. *Environ. Sci. Technol.* **2009**, *43*, 2362–2367. [CrossRef]
82. Lu, A.H.; Salabas, E.L.; Schüth, F. Magnetic nanoparticles: Synthesis, protection, functionalization, and application. *Angew. Chem. Int. Ed.* **2007**, *46*, 1222–1244. [CrossRef]
83. Yu, C.; Geng, J.; Zhuang, Y.; Zhao, J.; Chu, L.; Luo, X. Preparation of the chitosan grafted poly (quaternary ammonium)/Fe_3O_4 nanoparticles and its adsorption performance for food yellow 3. *Carbohydr. Polym.* **2016**, *152*, 327–336. [CrossRef]
84. Sayana, K.V.; Prajwal, K.; Deeksha, K.J.; Vishalakshi, B.; Vishwanath, T. Magnetized CNTs incorporated MBA cross-linked guar gum nano-composite for methylene blue dye removal. *J. Appl. Polym. Sci.* **2024**, *141*, e54868. [CrossRef]
85. Ege, K.; Arzum, Ç.; Alattin, Ç.; Elif, A. Enhanced photocatalytic dye degradation using surface-modified tungstophosphoric acid-iron magnetic nanocatalyst. *Opt. Mater.* **2024**, *148*, 114816. [CrossRef]
86. Zhao, C.; Wang, X.; Zhang, S.; Sun, N.; Zhou, H.; Wang, G.; Zhang, Y.; Zhang, H.; Zhao, H. Porous carbon nanosheets functionalized with Fe_3O_4 nanoparticles for capacitive removal of heavy metal ions from water. *Environ. Sci. Water Res. Technol.* **2020**, *6*, 331–340. [CrossRef]
87. Yang, W.; Chen, H.; Han, X.; Ding, S.; Shan, Y.; Liu, Y. Preparation of magnetic Co-Fe modified porous carbon from agricultural wastes by microwave and steam activation for mercury removal. *J. Hazard. Mater.* **2020**, *5*, 120981. [CrossRef] [PubMed]

88. Elessawy, N.A.; El-Sayed, E.M.; Ali, S.; Elkady, M.F.; Elnouby, M.; Hamad, H.A. One-pot green synthesis of magnetic fullerene nanocomposite for adsorption characteristics. *J. Water Process Eng.* **2020**, *34*, 101047. [CrossRef]
89. Elessawy, N.A.; Elnouby, M.; Gouda, M.H.; Hamad, H.A.; Taha, N.A.; Gouda, M.; Eldin, M.S.M. Ciprofloxacin removal using magnetic fullerene nanocomposite obtained from sustainable PET bottle wastes: Adsorption process optimization, kinetics, isotherm, regeneration and recycling studies. *Chemosphere* **2020**, *239*, 124728. [CrossRef]
90. Deng, Y.; Ok, Y.S.; Mohan, D.; Pittman, C.U.; Dou, X. Carbamazepine removal from water by carbon dot-modified magnetic carbon nanotubes. *Environ. Res.* **2019**, *169*, 434–444. [CrossRef]
91. Sun, A.C. Synthesis of magnetic carbon nanodots for recyclable photocatalytic degradation of organic compounds in visible light. *Adv. Powder Technol.* **2018**, *29*, 719–725. [CrossRef]
92. Zhang, H.; Chen, S.; Zhang, H.; Fan, X.; Gao, C.; Yu, H.; Quan, X. Carbon nanotubes-incorporated MIL-88B-Fe as highly efficient Fenton-like catalyst for degradation of organic pollutants. *Front. Environ. Sci. Eng.* **2019**, *13*, 18. [CrossRef]
93. Zhang, Z.; Chen, H.; Wu, W.; Pang, W.; Yan, G. Efficient removal of Alizarin Red S from aqueous solution by polyethyleneimine functionalized magnetic carbon nanotubes. *Bioresour. Technol.* **2019**, *293*, 122100. [CrossRef]
94. Salam, M.A.; El-Shishtawy, R.M.; Obaid, A.Y. Synthesis of magnetic multi-walled carbon nanotubes/magnetite/chitin magnetic nanocomposite for the removal of Rose Bengal from real and model solution. *J. Ind. Eng. Chem.* **2014**, *20*, 3559–3567. [CrossRef]
95. Cheng, J.; Chang, P.R.; Zheng, P.; Ma, X. Characterization of magnetic carbon nanotube-cyclodextrin composite and its adsorption of dye. *Ind. Eng. Chem. Res.* **2014**, *53*, 1415–1421. [CrossRef]
96. Zhu, H.; Fu, Y.; Jiang, R.; Yao, J.; Liu, L.; Chen, Y.; Xiao, L.; Zeng, G. Preparation, characterization and adsorption properties of chitosan modified magnetic graphitized multi-walled carbon nanotubes for highly effective removal of a carcinogenic dye from aqueous solution. *Appl. Surf. Sci.* **2013**, *285*, 865–873. [CrossRef]
97. Gao, H.; Zhao, S.; Cheng, X.; Wang, X.; Zheng, L. Removal of anionic azo dyes from aqueous solution using magnetic polymer multi-wall carbon nanotube nanocomposite as adsorbent. *Chem. Eng. J.* **2013**, *223*, 84–90. [CrossRef]
98. Madrakian, T.; Afkhami, A.; Ahmadi, M.; Bagheri, H. Removal of some cationic dyes from aqueous solutions using magnetic-modified multi-walled carbon nanotubes. *J. Hazard. Mater.* **2011**, *196*, 109–114. [CrossRef]
99. Yan, L.; Chang, P.R.; Zheng, P.; Ma, X. Characterization of magnetic guar gum-grafted carbon nanotubes and the adsorption of the dyes. *Carbohydr. Polym.* **2012**, *87*, 1919–1924. [CrossRef]
100. Qu, S.; Huang, F.; Yu, S.; Chen, G.; Kong, J. Magnetic removal of dyes from aqueous solution using multi-walled carbon nanotubes filled with Fe_2O_3 particles. *J. Hazard. Mater.* **2008**, *160*, 643–647. [CrossRef]
101. Ranjbar, E.; Ghiassi, R.; Baghdadi, M.; Ruhl, A.S. Bisphenol A removal in treated wastewater matrix at neutral pH using magnetic graphite intercalation compounds as persulfate activators. *Water Environ. Res.* **2023**, *95*, e10835. [CrossRef]
102. Ruan, C.P.; Ai, K.L.; Lu, L.H. An Acid-resistant Magnetic Co/C Nanocomposite for Adsorption and Separation of Organic Contaminants from Water. *Chin. J. Anal. Chem.* **2016**, *44*, 224–231. [CrossRef]
103. Wang, L.; Zhao, Q.; Hou, J.; Yan, J.; Zhang, F.; Zhao, J.; Ding, H.; Li, Y.; Ding, L. One-step solvothermal synthesis of magnetic Fe_3O_4-graphite composite for Fenton-like degradation of levofloxacin. *J. Environ. Sci. Health Part A Toxic/Hazard. Subst. Environ. Eng.* **2016**, *51*, 52–62. [CrossRef]
104. Bharath, G.; Alhseinat, E.; Ponpandian, N.; Khan, M.A.; Siddiqui, M.R.; Ahmed, F.; Alsharaeh, E.H. Development of adsorption and electrosorption techniques for removal of organic and inorganic pollutants from wastewater using novel magnetite/porous graphene-based nanocomposites. *Sep. Purif. Technol.* **2017**, *188*, 206–218. [CrossRef]
105. Yu, B.; Zhang, X.; Xie, J.; Wu, R.; Liu, X.; Li, H.; Chen, F.; Yang, H.; Ming, Z.; Yang, S.-T. Magnetic graphene sponge for the removal of methylene blue. *Appl. Surf. Sci.* **2015**, *351*, 765–771. [CrossRef]
106. Zhao, G.; Mo, Z.; Zhang, P.; Wang, B.; Zhu, X.; Guo, R. Synthesis of graphene/Fe_3O_4/NiO magnetic nanocomposites and its application in photocatalytic degradation the organic pollutants in wastewater. *J. Porous Mater.* **2015**, *22*, 1245–1253. [CrossRef]
107. Yang, X.; Li, J.; Wen, T.; Ren, X.; Huang, Y.; Wang, X. Adsorption of naphthalene and its derivatives on magnetic graphene composites and the mechanism investigation. *Colloids Surf. A Physicochem. Eng. Asp.* **2013**, *422*, 118–125. [CrossRef]
108. Yao, Y.; Miao, S.; Liu, S.; Ma, L.P.; Sun, H.; Wang, S. Synthesis, characterization, and adsorption properties of magnetic Fe_3O $_4$@graphene nanocomposite. *Chem. Eng. J.* **2012**, *184*, 326–332. [CrossRef]
109. Li, N.; Zheng, M.; Chang, X.; Ji, G.; Lu, H.; Xue, L.; Pan, L.; Cao, J. Preparation of magnetic $CoFe_2O_4$-functionalized graphene sheets via a facile hydrothermal method and their adsorption properties. *J. Solid State Chem.* **2011**, *184*, 953–958. [CrossRef]
110. Wang, C.; Feng, C.; Gao, Y.; Ma, X.; Wu, Q.; Wang, Z. Preparation of a graphene-based magnetic nanocomposite for the removal of an organic dye from aqueous solution. *Chem. Eng. J.* **2011**, *173*, 92–97. [CrossRef]
111. Islam, M.R.; Ferdous, M.; Sujan, M.I.; Mao, X.; Zeng, H.; Azam, M.S. Recyclable Ag-decorated highly carbonaceous magnetic nanocomposites for the removal of organic pollutants. *J. Colloid Interface Sci.* **2020**, *562*, 52–62. [CrossRef]
112. Chang, S.; Zhang, Q.; Lu, Y.; Wu, S.; Wang, W. High-efficiency and selective adsorption of organic pollutants by magnetic $CoFe_2O_4$/graphene oxide adsorbents: Experimental and molecular dynamics simulation study. *Sep. Purif. Technol.* **2020**, *238*, 119400. [CrossRef]
113. Chen, Z.; Zheng, Y.; Liu, Y.; Zhang, W.; Wang, Y.; Guo, X.; Tang, X.; Zhang, Y.; Wang, Z.; Zhang, T. Magnetic Mn-Doped Fe_3O_4 hollow Microsphere/RGO heterogeneous Photo-Fenton Catalyst for high efficiency degradation of organic pollutant at neutral pH. *Mater. Chem. Phys.* **2019**, *238*, 121893. [CrossRef]

114. Mishra, A. Study of organic pollutant removal capacity for magnetite@ graphene oxide nanocomposites. *Vacuum* **2018**, *15* 524–529. [CrossRef]
115. Bai, S.; Shen, X.; Zhong, X.; Liu, Y.; Zhu, G.; Xu, X.; Chen, K. One-pot solvothermal preparation of magnetic reduced graphen oxide-ferrite hybrids for organic dye removal. *Carbon* **2012**, *50*, 2337–2346. [CrossRef]
116. Song, Z.J.; Ran, W.; Wei, F.Y. One-step approach for the synthesis of $CoFe_2O_4$ @rGO core-shell nanocomposites as efficien adsorbent for removal of organic pollutants. *Water Sci. Technol.* **2017**, *75*, 397–405. [CrossRef]

surfaces

MDPI

Article

Synthesis of Granular Free-Binder ZSM-5 Zeolites Using Different Amorphous Aluminosilicates

Alina Kh. Ishkildina [1], Olga S. Travkina [1], Dmitry V. Serebrennikov [1,*], Rufina A. Zilberg [2], Artur I. Malunov [1], Nadezhda A. Filippova [1], Boris I. Kutepov [1] and Marat R. Agliullin [1]

1 Institute of Petrochemistry and Catalysis, Ufa Federal Research Centre of the Russian Academy of Sciences (UFRC RAS), 450075 Ufa, Russia; simchanka@mail.ru (O.S.T.); malunovv@mail.ru (A.I.M.); filippova-ink@yandex.ru (N.A.F.)

2 Department of Analytical Chemistry, Ufa University of Science and Technology, 450076 Ufa, Russia; zilbergra@yandex.ru

* Correspondence: d25c25@yandex.ru

Abstract: In this paper, we discuss options for the synthesis of granular free-binder ZSM-5 zeolites using synthetic aluminosilicates prepared by sol-gel technology with organic and inorganic silicon sources. It has been shown that the properties of the amorphous aluminosilicate used to prepare the initial granules influence the crystallization conditions, as well as the morphology and size of the crystals formed from granular ZSM-5 zeolite. The granular free-binder Pt/ZSM-5 with a developed secondary porous structure showed higher activity in the hydrocracking of hexadecane than the granular binder Pt/ZSM-5. At a reaction temperature of 220 °C, the conversion of n-hexadecane in the granular free-binder sample was 59.1%. At the same time, the selectivity for hexadecane isomers was 15.7%.

Keywords: ZSM-5 zeolite; amorphous aluminosilicate; secondary porosity; granular zeolite; hydrocracking of n-paraffins

1. Introduction

Currently, zeolites are widely used in the production of modern adsorbents and catalysts for the petrochemical and oil refining industries due to their successful combination of microporous structures, strong acid sites, molecular sieving effects, and high hydrothermal and thermal stability [1,2]. Among the various types of molecular sieves, ZSM-5 zeolite is the most widely used in catalytic systems. It belongs to the MFI structural type and has a three-dimensional porous structure consisting of rectangular and elliptical channels with dimensions of 5.3 × 5.6 Å and 5.1 × 5.5 Å, respectively [3].

The main efforts of researchers in the field of ZSM-5 zeolite synthesis have focused on the development of methods for the preparation of nanoscale and hierarchical crystals in order to overcome the diffusion limitations in their micropores [4–10]. It has been demonstrated that catalytic systems based on nanoscale and hierarchical ZSM-5 zeolite crystal structures exhibit high activity, selectivity, and stability in several industrially important catalytic processes [11–15].

One of these important processes is the hydroconversion of n-alkanes to produce high-quality fuel and oil. This process proceeds in a hydrogen medium by a bifunctional mechanism that requires a catalyst, which is an acid support on which acid center isomerization and cracking take place, with metal nanoparticles (usually Pt or Pd) providing the hydrogenation-dehydrogenation stages [16,17].

The selectivity of ZSM-5 for the hydroisomerization of n-alkanes is rather low compared to that of other zeolites whose micropores consist of 10R rings, such as MCM-22 and ZSM-23 [18,19]; the main reaction in this case is hydrocracking. However, catalysts based on ZSM-5 zeolite have been successfully used in the industrial process of dewaxing diesel fuel to improve its low-temperature properties [16].

It has been shown [20] that on ZSM-5 zeolite with a developed secondary porous structure, it is possible to achieve higher activity values in the conversion of hexadecane and a slight increase in selectivity for hexadecane isomers compared to conventional microporous ZSM-5 due to the reduction of the diffusion path length.

Unfortunately, most methods for synthesizing these materials rely on the use of crystal growth modifiers and pore-forming templates. However, the main drawback of these methods for producing nanoscale and hierarchical zeolite crystals, such as ZSM-5 zeolite, is their high cost, which makes them unsuitable for large-scale industrial production.

In industrial processes, zeolite-based catalysts, such as ZSM-5, are used in granular form. These granules are produced by mixing and pelletizing powdered zeolite with boehmite, followed by drying at temperatures between 120 and 150 °C, and calcining at temperatures between 500 and 650 °C. During this process, boehmite is transformed into γ-Al_2O_3 [21,22]. During granulation, the pores of the zeolite crystals may be partially blocked by a binder. The amount of binder used depends on the amount of Al_2O_3 present in the mixture [21,22].

In [23], a method for the preparation of granular ZSM-5 zeolite without the use of binders was proposed. This method involves crystallizing the granules at 180 °C for 24 h in a 0.01 M NaOH solution, using dried aluminosilicate gel granules prepared with sodium silicate as a temporary binder. It has been shown that this approach allows for the synthesis of granular ZSM-5 zeolite without the use of binders. However, it should be noted that this study does not provide information on the mechanical strength of the granules, which is a critical parameter for industrial catalysts.

In [24], a method was proposed for the synthesis of granular zeolite Na-Y with a developed secondary porous structure. The method involves the crystallization of granules consisting of Y zeolite and metakaolin (SiO_2/Al_2O_3 = 2.0–2.21), which, during the crystallization process, are transformed into individual clusters of crystals, including nanoscale crystals. This approach has made it possible to obtain granular materials with a high degree of crystallinity without binders and with a developed secondary (micro-, meso-, and macroporous) porous structure. However, the use of kaolin for the preparation of zeolites with a molar SiO_2/Al_2O_3 ratio higher than 30 is not promising due to its low silicon content.

Therefore, the aim of this study was to develop a method for the synthesis of granular binder-free ZSM-5 zeolites using amorphous aluminosilicates obtained by the sol-gel process as a temporary binder.

2. Materials and Methods

2.1. Materials

The following reagents were used in this work: sodium aluminate ($Na_2O(Al_2O_3)_x$ xH_2O, 99%, No. CAS 11138-49-1) and ground silica gel (SiO_2, No. CAS 112926-00-8), tetrabutylammonium bromide (TBABr, No. CAS 1643-19-2), ethanol (C_2H_5OH, 99%, No. CAS 64-17-5), tetraethyl orthosilicate (TEOS, 99%, No. CAS 78-10-4), aluminum nitrate ($Al(NO_3)_3$*$9H_2O$, 98%, No. CAS 7784-27-2), ammonia (NH_4OH, 30%, No. CAS 1336-21-6), polyvinyl alcohol (PVA, 99%, No. CAS 9002-89-5), ammonium nitrate (NH_4NO_3, 98%, No. CAS 6484-52-2), $H_2PtCl_6 \times 6H_2O$ (99%, № CAS 26023-84-7) purchased from Sigma-Aldrich

(Saint Louis, MO, USA); sodium silicate ($Na_2O(SiO_2)_x$ xH_2O, 99%, No. CAS 6834-92-0), aluminum sulfate ($Al_2(SO_4)_3$*$18H_2O$, 99%, No. CAS 7784-31-8), n-hexadecane (H-$C_{16}H_{34}$, 99%, No. CAS 26023-84-7) purchased from Acros Organics (Geel, Belgium); boehmite (AlO(OH), 78% Al_2O_3, No. CAS 1318-23-6) purchased from Sasol (Hamburg, Germany); nitric acid (HNO_3, 67%, No. CAS 7697-37-2) purchased from Reachem (Moscow, Russia).

2.2. *Synthesis of Powdered ZSM-5 Zeolite*

Highly dispersed Na-ZSM-5 zeolite with a SiO_2/Al_2O_3 ratio of 50 was obtained by hydrothermal crystallization in Teflon-coated autoclaves at 160 °C for 48 h from an amorphous alkaline aluminosilicate with the following composition: $0.03Na_2O \cdot 0.04TBABr \cdot 0.02Al_2O_3 \cdot 1.00SiO_2 \cdot 16.00H_2O$, prepared by mixing solutions of sodium silicate and sodium aluminate, ground silica gel, and an organic template—tetrabutylammonium bromide.

The crystallization products were separated by centrifugation and washed with distilled water until the pH was neutral. They were then dried at 150 °C for 8–10 h. The resulting powdered zeolite was named the Na-ZSM-5 sample.

2.3. *Synthesis of Amorphous Aluminosilicates*

Amorphous aluminosilicate (molar ratio of SiO_2/Al_2O_3 = 50) was synthesized using an organic silicon source by a two-step sol-gel method following the procedure described in [25]. In the first stage, calculated amounts of distilled water and ethanol were added to tetraethyl orthosilicate under intensive stirring. Aluminum nitrate was then added to the resulting solution. The resulting solution, with a pH of approximately 3, was then kept in a thermostat at 60 °C for 25 h until the gelation point was reached. An aqueous solution of ammonia was then added to the resulting gel with vigorous stirring until a pH of 10 was reached. The mixture was then aged at 25 °C for a further 24 h. Finally, the gel was dried at 120 °C for 5 h to obtain the final product. The resulting amorphous aluminosilicate was named the ASM-1 sample.

An amorphous aluminosilicate with a molar ratio of SiO_2/Al_2O_3 = 50 was synthesized using an inorganic silicon source. The synthesis process involved mixing aqueous solutions of sodium silicate and aluminum sulfate to form a gel. The gel was then aged for 24 h at 25 °C, filtered, and washed with distilled water to remove impurities. After this washing step, the amorphous aluminosilicate was dried at 120 °C for 5 h. The final product of this process, called the ASM-2 sample, is an amorphous material with the desired molar ratio.

2.4. *Preparation of Granules Containing Zeolite ZSM-5 and Amorphous Aluminosilicates*

The initial granules were prepared by combining powdered Na-ZSM-5 zeolite and amorphous aluminosilicates in a mixer (VINCI Technologies MX 0.4). The resulting mixture was moistened with a solution of polyvinyl alcohol to form granules with a diameter of 1.4–1.6 mm and a length of 5–6 mm. The granules were then formed on an extruder (VINCI Technologies VTE1).

The contents of powdered ZSM-5 zeolite and amorphous aluminosilicate in the granules were 60% and 40% by weight, respectively. The granules obtained using amorphous aluminosilicates ASM-1 and ASM-2 were designated as Na-ZSM-5-ASM-1 and Na-ZSM-5-ASM-2 zeolites, respectively.

2.5. *Preparation of Granules Containing ZSM-5 Zeolite and Boehmite*

Granular ZSM-5 zeolite in the Na form, with a binder, was prepared as follows: The Na-ZSM-5 sample and boehmite were thoroughly mixed in a mixer to obtain a homogeneous mixture. The content of boehmite in the granules, expressed as Al_2O_3, was 30% by weight.

The resulting mixture was wetted with a 5% nitric acid solution and extruded into granules (diameter: 1.4–1.6 mm, length: 5–6 mm) using an extruder. The granules were then dried at 150 °C for 24 h and calcined at 500–600 °C for 6 h to convert boehmite to γ-Al_2O_3. The sample obtained by this method was designated as Na-ZSM-5-BD.

2.6. Preparation of Granular Binder-Free ZSM-5 Zeolites

Granular Na-ZSM-5-ASM-1 and Na-ZSM-5-ASM-2 zeolites were crystallized in a solution of sodium silicate. The concentrations of sodium and silicon were chosen on the basis of the composition of the reaction mixture (RM): 2.3R: $3.2Na_2O$: Al_2O_3: $60SiO_2$: $550H_2O$ (R—organic template, tetrabutylammonium bromide).

The granules were stored at room temperature for 1–32 h prior to crystallization. Crystallization was performed in a Teflon-lined autoclave. The temperature for crystallization was 160 °C, and the duration was 48 h. After the crystallization process, the samples were washed to remove any residual components from the mother liquor, dried at 120 °C for 5–6 h, and finally calcined at 600 °C for 3–4 h to remove the template.

The samples obtained by crystallization of Na-ZSM-5-ASM-1 and Na-ZSM-5-ASM-2 were designated as Na-ZSM-5-WB-1 and Na-ZSM-5-WB-2, respectively.

2.7. Preparation of H-Form Zeolites

The H-form of the powder and granular samples was obtained after crystallization by ion exchange of Na^+ cations with NH_4^+ cations in an aqueous solution of ammonium nitrate at 70 °C for 1 h with stirring. The granules were then dried at 120 °C and calcined at 550 °C for 4 h in air. The ion exchange process was performed three times. The residual sodium (Na_2O) content was less than 0.003%. The H-form samples were assigned an H-index.

2.8. Preparation of a Bifunctional Catalyst

Fractions of 0.1 to 0.5 mm were obtained from H-ZSM-5-WB-1, H-ZSM-5-WB-2, and H-ZSM-5-BD samples by grinding and passing through sieves. This material was then heat-treated at 350 °C for 6 h in air. The catalyst was then impregnated with an aqueous solution of $H_2PtCl_6 \times 6H_2O$ at a concentration of 0.5 wt. % of Pt per support weight. The catalyst was then dried at 100 °C for 24 h and calcined at 550 °C for 5 h. Prior to the reaction, the catalyst was reduced under a hydrogen atmosphere at 400 °C for 5 h. Samples containing 0.5% Pt were designated as Pt/ZSM-5-WB-1, Pt/ZSM-5-WB-2, and Pt/ZSM-5-BD.

2.9. Zeolite Research Methods

The chemical composition of the samples was determined using a Shimadzu EDX-7000P spectrometer manufactured by Shimadzu Corporation.

The phase composition and crystallinity of the samples were determined by X-ray diffraction using a Shimadzu XRD 7000 diffractometer with CuKα radiation. Scanning was performed in the range of angles 2θ from 5° to 40° with increments of 1°/min. The X-ray images were analyzed using the Shimadzu PC XRD software (version 7.04) and the PDF2 database (version 2.2201). Crystallinity was calculated using the Shimadzu Crystallinity software (version 7.04), taking into account the halo in the range of 15–30°, which is characteristic of the amorphous phase.

The morphology and crystal size of the samples were analyzed by scanning electron microscopy (SEM) using a Hitachi Regulus SU8220 microscope. Images were taken in the secondary electron mode with an acceleration voltage of 5 kV. Prior to imaging, the samples were positioned on a 25 mm diameter aluminum stage and secured with conductive carbon tape.

A target-oriented approach was used to optimize the analytical measurements [26]. Prior to the measurements, the samples were deposited on 3 mm carbon-coated copper grids from an isopropanol suspension. Observations were carried out using a Hitachi Regulus8230 field-emission scanning electron microscope (FE-SEM). Images were taken in the transmitted electron mode at an acceleration voltage of 30 kV.

The characteristics of the porous structure, including the BET-specific surface area and the volume of micro-, meso-, and macropores, were measured using low-temperature N_2 adsorption−desorption on a Quantachrome Nova 1200e sorption meter and mercury porosimetry using a Carlo Erba Porosimeter-2000.

The volume of micropores in the presence of mesopores was calculated using the t-plot method. The pore size distribution was determined using the BJH model (Halenda). Mercury porosimetry was used to determine the volume of the macropores. Mercury penetration into pores with radii ranging from 30 to 10,000 Å was carried out under pressures ranging from 0.1 to 200 MPa. During data processing, differential curves were obtained for the distribution of pores over radii, from which the contributions of pores of different sizes to the total volume of the porous space of the catalyst were determined.

The acid site types and concentrations were determined by IR spectroscopy after pyridine adsorption (IR-Py). IR spectra of the adsorbed pyridine were recorded using a Bruker Vertex-70V IR spectrometer (Bruker Optic GmbH, Ettlingen, Germany) with a resolution of 4 cm^{-1}. The samples were precalcined at 450 °C for 2 h in a 10^{-2} Pa vacuum. Pyridine was adsorbed onto the molecular sieve samples at 150 °C for 30 min. Pyridine was desorbed at 150 °C, 250 °C, and 350 °C. The number of Brønsted and Lewis acid sites was determined by integrating the bands in the ranges 1570–1510 and 1475–1410, respectively. The average molar extinction coefficients from references [27] were used to calculate the concentrations.

The mechanical strength of the granular samples was tested on a LinteL PC-21 machine under static conditions using the compression method. Cylindrical granules with a length of 5–6 mm and a diameter of 1.4–1.6 mm were tested according to the ASTM D6175 standard.

2.10. Methods for Testing Catalysts

The catalytic transformations of n-hexadecane were studied in a flow reactor at temperatures between 220 and 260 °C and a pressure of 3 MPa. A molar ratio of $H_2/\text{H-}C_{16}H_{34} = 10$ was used, and the mass feed rate was 2 h^{-1}. The reaction products were analyzed by gas-liquid chromatography on a Chromatek-Crystal 5000 chromatograph with a flame ionization detector and a glass capillary column (50 m, HP-1). Chromato-mass spectrometry was also used with a Shimadzu instrument whose chromatograph was equipped with a DB-5 column (50 m).

3. Results and Discussion

The catalytic properties of ZSM-5 zeolite depended on the degree of crystallinity and the presence of impurity phases. Figure 1 shows the X-ray diffraction patterns of amorphous aluminosilicates, initial powdered ZSM-5 zeolite, granules containing amorphous aluminosilicates and boehmite, and granules of ZSM-5 zeolite without a binder.

It can be seen from the X-ray diffraction pattern of the Na-ZSM-5 zeolite powder sample (Figure 1c) that there are several intense signals at specific angles: 7.92°, 8.94°, 23.04°, 23.38°, and 23.94° degrees 2θ. These specific angles are characteristic of the MFI crystal structure with high phase purity (PDF№00-037-0390).

Amorphous aluminosilicates (Figure 1a,b) show a strong diffraction peak in the 2θ angle range of 15–30°, regardless of the silicon source used. Granular samples containing ZSM-5 zeolite and amorphous aluminosilicates (Figure 1d,e) show a well-defined diffraction peak in the 20–30° angle range, with a degree of crystallinity that does not exceed 70% for these granules.

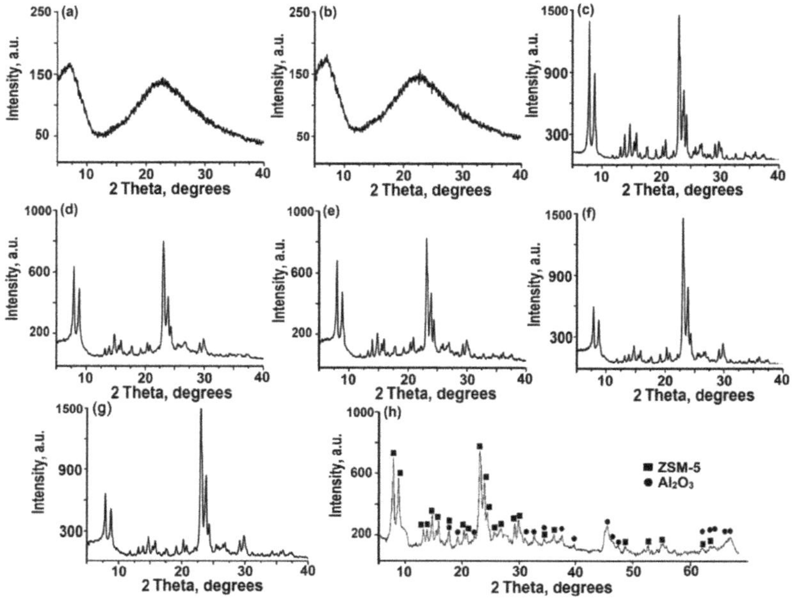

Figure 1. X-ray diffraction patterns of powdered and granular samples: (**a**)—ASM-1; (**b**)—ASM-2; (**c**)—Na-ZSM-5; (**d**)—Na-ZSM-5-ASM-1; (**e**)—Na-ZSM-5-ASM-2; (**f**)—Na-ZSM-5-WB-1; (**g**)—Na-ZSM-5-WB-2; and (**h**)—Na-ZSM-5-BD.

After crystallization, the amorphous phase in the granules transforms into ZSM-5 zeolite with a high degree of purity and a crystallinity of at least 95% (Figure 1f,g). These results indicate that the granular sample has a similar level of crystallinity to that of powdered ZSM-5 zeolite.

Table 1 shows the results of the study of the chemical composition of amorphous aluminosilicates and powdered and granular ZSM-5 zeolites. The SiO_2/Al_2O_3 ratio in powdered Na-ZSM-5 zeolite was found to be slightly lower than that in the reaction mixture due to incomplete aluminum incorporation during crystallization. This is in contrast to amorphous aluminosilicates, where the SiO_2/Al_2O_3 ratio was closer to the calculated value.

In granular samples, the SiO_2/Al_2O_3 ratios are also similar before and after crystallization, indicating that the SiO_2/Al_2O_3 ratio of amorphous aluminosilicates and zeolite ZSM-5 crystals remains constant during synthesis. However, in the granular sample of Na-ZSM-5-BD zeolite, there is a decrease in the SiO_2/Al_2O_3 ratio due to the addition of boehmite to the granules.

The morphology and size of zeolite crystals is also one of the key factors affecting their catalytic properties [28]. Figure 2 shows scanning transmission electron microscope (STEM) images of amorphous aluminum silicates obtained using different silicon sources. These images show that amorphous aluminosilicates are xerogels composed of spherical particles. ASM-1 aluminosilicate, obtained using TEOS, has particles with sizes between 5 and 10 nm, whereas ASM-2, obtained using an inorganic silicon source, is characterized by larger particles with sizes between 10 and 20 nm.

Table 1. Chemical composition of amorphous aluminosilicates, powdered, and granular ZSM-5 zeolites.

Sample	Chemical Composition, wt. %			SiO_2/Al_2O_3 Molar Ratio
	Na_2O	Al_2O_3	SiO_2	
ASM-1	0	5.6	94.4	29
ASM-2	3.2	5.3	91.5	29
Na-ZSM-5	2.0	3.3	94.7	49
Na-ZSM-5-ASM-1	1.2	4.2	94.6	38
Na-ZSM-5-ASM-2	2.5	4.0	93.5	39
Na-ZSM-5-WB-1	2.0	3.4	94.6	47
Na-ZSM-5-WB-2	2.0	3.4	94.6	47
Na-ZSM-5-BD	1.5	22.7	75.8	6

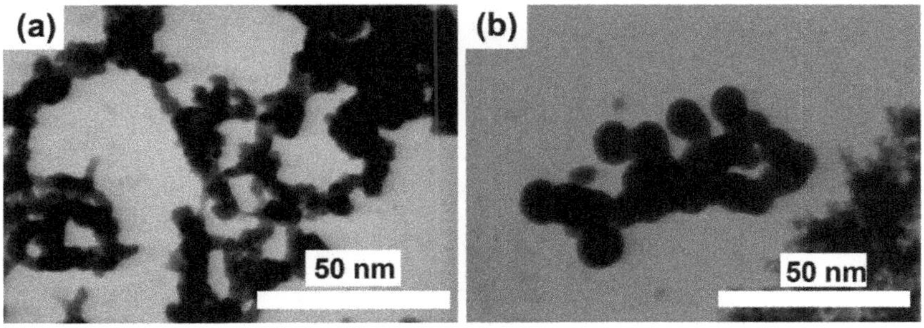

Figure 2. STEM images of samples of amorphous aluminosilicates: (**a**)—ASM-1 and (**b**)—ASM-2.

According to the SEM data (Figure 3), the powdered Na-ZSM-5 sample consists of crystals in the form of fused prisms with sizes ranging from 300 to 500 nm. The Na-ZSM-5-ASM-1 and Na-ZSM-5-ASM-2 granules produced from the Na-ZSM-5 sample are composite materials composed of Na-ZSM-5 molecular sieve crystals with sizes ranging from 300 to 500 nm and highly dispersed particles of amorphous aluminosilicate filling the spaces between the crystals. The granules of the Na-ZSM-5-WB-1 and Na-ZSM-5-WB-2 zeolite samples are composed of clusters of crystals with different morphologies. These crystals can be cubic in shape with sizes between 50 and 200 nm or elongated prismatic in shape with sizes between 300 and 500 nm. It is worth noting that the Na-ZSM-5-WB-1 zeolite sample contains crystals that are smaller in size compared to those in the Na-ZSM-5-WB-2 zeolite sample. The results obtained can be explained by the fact that during the crystallization of granules containing ASM-1 aluminosilicate, a higher degree of supersaturation of crystal nuclei is created compared to the crystallization of granules containing ASM-2 aluminosilicate. This results in the formation of more finely dispersed crystals.

The Na-ZSM-5-BD zeolite sample consists of ZSM-5 molecular sieve crystals, with crystal sizes ranging from 300 to 500 nm, and highly dispersed aluminum oxide particles filling the spaces between the crystals.

Figure 4 shows photographs of the granules both before and after the crystallization process. It is clear that crystallization had little or no effect on the appearance or size of the granules.

Granular industrial catalysts based on ZSM-5 zeolite require high mechanical strength, as large amounts of catalyst in the reactor can cause the lower layers to collapse under the pressure of their own weight.

Figure 3. SEM images of samples of powdered and granular ZSM-5 zeolites: (**a**)—Na-ZSM-5; (**b**)—Na-ZSM-5-ASM-1; (**c**)—Na-ZSM-5-ASM-2; (**d**)—Na-ZSM-5-WB-1; (**e**)—Na-ZSM-5-WB-2; and (**f**)—Na-ZSM-5-BD.

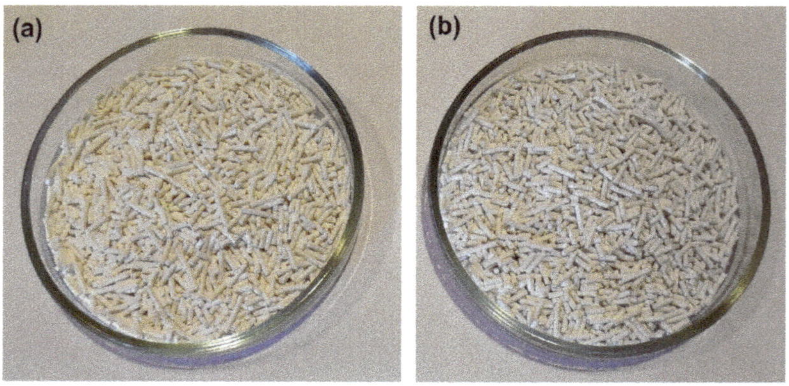

Figure 4. Photographs of granules before and after crystallization: (**a**)—Na-ZSM-5-ASM-1; (**b**)—Na-ZSM-5-WB-1.

Table 2 presents the data on the mechanical strength of ZSM-5 zeolite granules after the crystallization process. The strength of the granules increased 2–3 times after crystallization compared to granules with amorphous aluminosilicates. This is due to the fusion of zeolite crystals from the initial granules with crystals formed during the crystallization of the amorphous component in the granules. As we mentioned previously, these granular materials form a single structure of fused crystals [24]. The Na-ZSM-5-WB-1 sample has a higher mechanical strength than the Na-ZSM-5-WB-2 sample due to the fact that its initial granules contain aluminosilicates with a more uniform particle distribution. This leads to an increase in the contact surface area and denser crystal fusion, resulting in an increase in strength.

Table 2. Mechanical strength of granular ZSM-5 zeolite samples.

Sample	Mechanical Crushing Strength (Radial)		Mechanical Crushing Strength (Uniaxial), N/mm^2
	N/Pellet	N/mm^2	
Na-ZSM-5-ASM-1	19 ± 4	4 ± 1	5 ± 1
Na-ZSM-5-ASM-2	17 ± 3	3 ± 1	4 ± 1
Na-ZSM-5-WB-1	60 ± 14	11 ± 3	12 ± 3
Na-ZSM-5-WB-2	58 ± 13	9 ± 2	10 ± 2
Na-ZSM-5-BD	43 ± 9	7 ± 1	8 ± 2

Results are shown as the mean ± standard deviation, SD (n = 24).

Comparing the mechanical strength of the Na-ZSM-5-WB and Na-ZSM-5-BD granules, it can be observed that the strength of the samples with a developed secondary porous structure is approximately 20% higher due to the unique formation process.

Figure 5 shows the nitrogen adsorption—desorption isotherms and pore size distribution for all samples. It can be seen that all isotherms are close to type IV, with a sharp increase in the low-pressure region and a hysteresis loop of type H3 between 0.8 and 1.0 pressure. This type of isotherm is typical for micro-mesoporous materials. The formation of mesopores in Na-ZSM-5 is due to the partial melting of nanoscale crystals into prism-like structures, as can be clearly seen from the SEM images (Figure 3).

Table 3 shows the characteristics of the porous structure of the powdered zeolite and the granules made from it. It can be seen that the introduction of ~40% by weight of amorphous aluminosilicate into the granules (Na-ZSM-5-ASM-1 zeolite sample) results in a decrease of about 30% in the volume of micropores and an increase in the volume of mesopores. This decrease in micropore volume and increase in mesopore volume is due to the addition of mesoporous amorphous aluminosilicate to the composition. The Na-ZSM-5-ASM-1 sample has a higher specific surface area and mesopore volume than the Na-ZSM-5-ASM-2 sample. This is because the former contains an aluminosilicate with a smaller particle size. After crystallization, granular samples experience an increase in micropore volume due to the formation of ZSM-5 zeolite crystals.

The specific surface area and volume of mesopores in the Na-ZSM-5-WB-1 sample are higher than those in the Na-ZSM-5-WB-2 sample due to the smaller crystals formed in its granules. In a granular sample containing a boehmite-based binder, there is a decrease in the specific surface area and volume of micropores due to the presence of the binder in the granules.

A higher volume of macropores was observed in the Na-ZSM-5-WB-2 zeolite sample than that in the Na-ZSM-5-WB-1 sample. This is due to the presence of larger crystals in the granules of the Na-ZSM-5-WB2 sample. Among the granular samples, the ZSM-5 samples without binders were characterized by the highest volume of macropores. Therefore, granular samples without binders can be considered as materials with secondary porosity, in which a well-developed secondary porous structure consisting of meso- and macropores has formed.

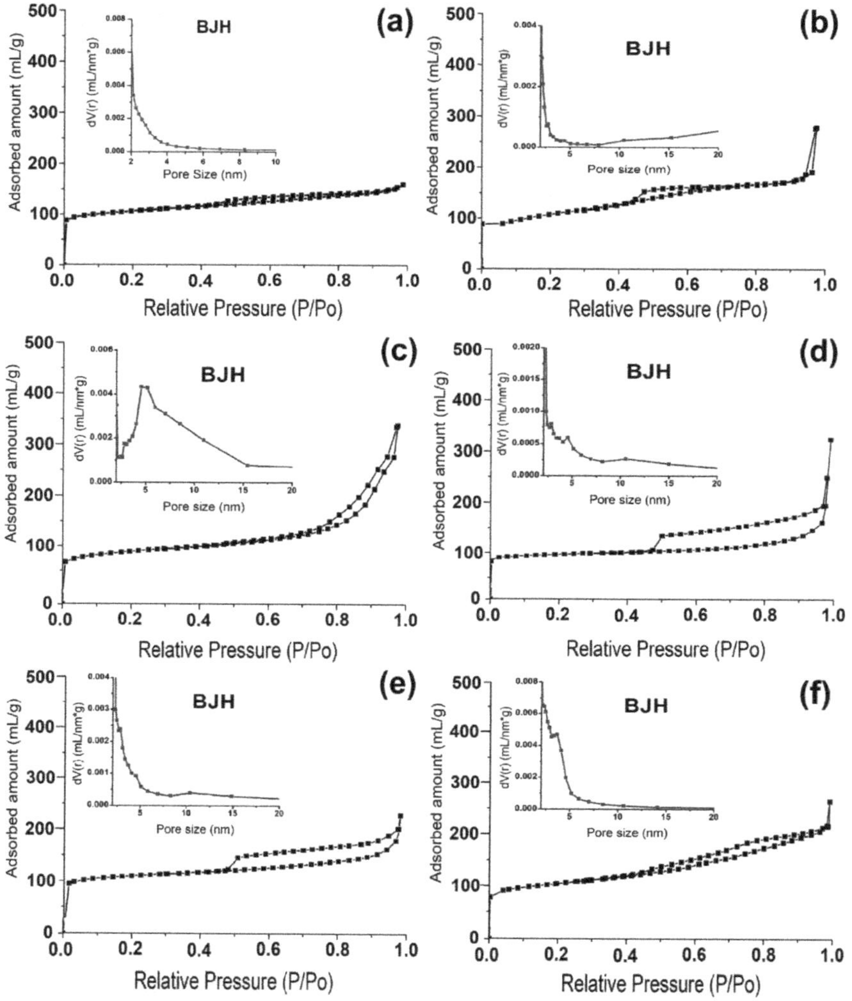

Figure 5. Nitrogen adsorption–desorption isotherms and pore size distribution of (BJH) for granular and powdered zeolite samples of ZSM-5: (**a**)—Na-ZSM-5; (**b**)—Na-ZSM-5-ASM-1; (**c**)—Na-ZSM-5-ASM-2; (**d**)—Na-ZSM-5-WB-1; (**e**)—Na-ZSM-5-WB-2; (**f**)—Na-ZSM-5-BD.

Table 3. Characteristics of the porous structure of granular and powdered ZSM-5 zeolite samples.

Sample	S_{BET}, m²/g	V_{micro}, cm³/g	V_{meso}, cm³/g	${}^{Hg}V_{\Sigma}$, cm³/g
Na-ZSM-5	285	0.12	0.01	-
Na-ZSM-5-ASM-1	363	0.06	0.37	0.35
Na-ZSM-5-ASM-2	353	0.06	0.30	0.30
Na-ZSM-5-WB-1	347	0.11	0.28	0.27
Na-ZSM-5-WB-2	329	0.11	0.20	0.21
Na-ZSM-5-BD	276	0.07	0.09	0.28

S_{BET}—specific surface area according to BET. V_{micro}—specific volume of micropores. V_{meso}—specific volume of mesopores. ${}^{Hg}V_{\Sigma}$—total pore volume (meso + macropores).

The acidic properties of the H-ZSM-5 zeolite samples were investigated by IR spectroscopy of pyridine adsorption (Figure 6). The bands at 1636 and 1546 cm^{-1} correspond to pyridine adsorbed on Brønsted acid sites (BAS). The bands at 1623 and 1454 cm^{-1} in H-ZSM-5-WB-1 (Figure 6b), H-ZSM-5-WB-2 (Figure 6c), and H-ZSM-5-BD-1 (Figure 6d) zeolite samples correspond to pyridine adsorbed on Lewis acid sites in zeolite (LAS). Samples H-ZSM-5-WB-1, H-ZSM-5-WB-2, and H-ZSM-5-BD show bands at 1614, 1601, and 1447 cm^{-1}, respectively, which are associated with adsorption on weak Lewis acid sites on extraframework aluminum. The band at 1490 cm^{-1} is attributed to both BAS and LAS.

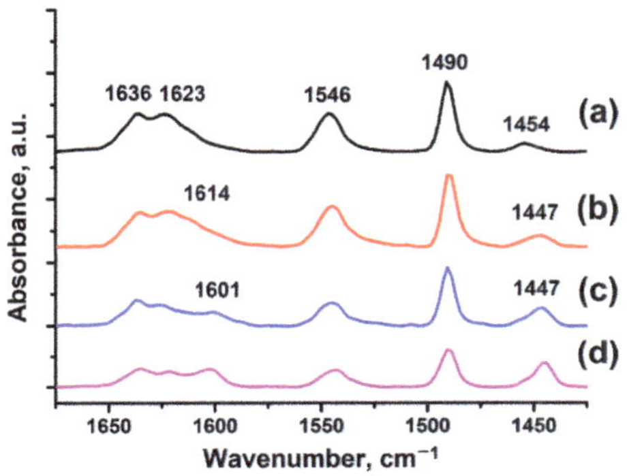

Figure 6. IR spectra of adsorbed pyridine for granular and powdered ZSM-5 zeolite samples: (**a**) H-ZSM-5, (**b**) H-ZSM-5-WB-1, (**c**) H-ZSM-5-WB-2, and (**d**) H-ZSM-5-BD.

The sample based on the original powdered ZSM-5 in H-form is mainly characterized by the presence of Brønsted acidity. Table 4 shows that the BAS content is 295 μmol/g, and the LAS content is 39 μmol/g. The total BAS concentration of the H-ZSM-5-WB-1 zeolite is higher than that of the original H-ZSM-5 powder at 318 μmol/g. However, the amount of strong BAS (desorption at 350 °C) is the same. This can be explained by the fact that the H-ZSM-5-WB-1 granular sample has a higher crystal dispersion-specific surface area and a higher concentration of acid sites.

Table 4. Concentrations of acid sites in granular and powdered ZSM-5 zeolite samples according to IR spectroscopy data with pyridine adsorption.

Sample	Acidity (μmol/g)					
	BAS			LAS		
	150 °C	250 °C	350 °C	150 °C	250 °C	350 °C
H-ZSM-5	295	256	223	39	25	24
H-ZSM-5-WB-1	318	278	224	59	27	16
H-ZSM-5-WB-2	206	173	132	90	34	15
H-ZSM-5-BD	148	124	99	106	39	17

The H-ZSM-5-WB-2 zeolite sample shows a decrease in the number of BAS to 206 μmol/g and an increase in the number of LAS to 90 μmol/g. The majority of these are weak Lewis acid sites (desorbing up to 250 °C). It is likely that there is more amorphous aluminosilicate present in the composition of H-ZSM-5-WB-2 than in H-ZSM-5-WB-1, as almost all of the amorphous component has crystallized in the latter.

Similarly, an even greater decrease in BAS concentration to 148 μmol/g in the H-ZSM-5-BD granular sample is associated with the dilution of the ZSM-5 zeolite by γ-Al_2O_3.

Apparently, γ-Al_2O_3 simultaneously acts as a source of weak LAS, and therefore, an increase in the LAS concentration to 106 μmol/g is observed.

As a result of testing the catalytic properties of Pt/ZSM-5 zeolite samples in the hydroconversion of n-hexadecane, we found that the highest activity was exhibited by the Pt/ZSM-5-WB-1 sample with the highest BAS concentration (Table 4) and specific surface area. The Pt/ZSM-5-BD sample with the lowest BAS concentration and specific surface area (Table 3) was the least active for hexadecane hydroconversion. At 220 °C, the hexadecane conversions were 59.1%, 50.2%, and 31.5% for the Pt/ZSM-5-WB-1, Pt/ZSM-5-WB-2, and Pt/ZSM-5-BD zeolite samples, respectively.

Although the Pt/ZSM-5-WB-1 and Pt/ZSM-5-WB-2 samples have the same SiO_2/Al_2O_3 values (Table 1), Pt/ZSM-5-WB-1 is characterized by a higher concentration of Brønsted acid sites, which play a significant role in the bifunctional mechanism of hydroconversion at the isomerization/cracking stage [20], and a higher specific surface area and thus better accessibility of acid sites for hexadecane molecules. Therefore, the granular binder-free Pt/ZSM-5-WB-1 sample is more active in the conversion of hexadecane. The Pt/ZSM-5-BD sample showed even lower activity because the binder in its composition led to a decrease in the content of H-ZSM-5 as the active component and probably a decrease in the availability of active sites due to partial pore blocking.

It has been found that the main reaction pathway for the conversion of hexadecane over Pt/ZSM-5 catalysts is hydrocracking (Figure 7), which produces hydrocarbons with carbon numbers ranging from 5 to 10 as the main products (Table 5). As the reaction temperature is increased from 220 to 260 °C, the yield of gaseous C_2–C_4 hydrocarbons increases, while the content of C_{11}–C_{13} hydrocarbons decreases. At the same time, there is a decrease in the concentration of normal-structure hydrocarbons relative to iso-structure hydrocarbons (i/n ratio).

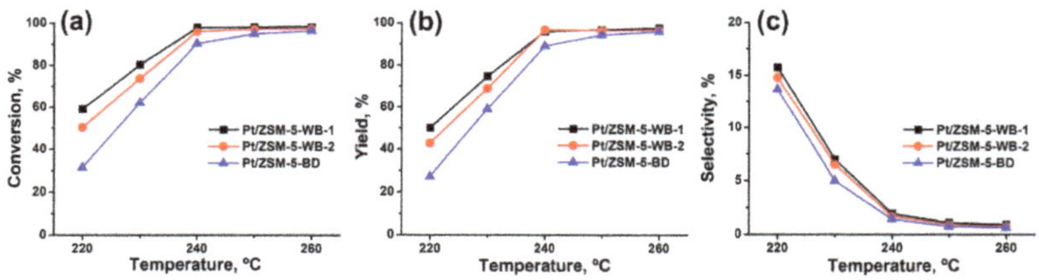

Figure 7. Catalytic transformations of n-hexadecane on Pt/ZSM-5 zeolite samples: (**a**)—Dependence of the conversion of n-hexadecane on the reaction temperature; (**b**)—Dependence of the yield of cracking products on the reaction temperature; (**c**)—Dependence of the selectivity for the formation of hexadecane isomers on the reaction temperature.

Another direction of hexadecane conversion using Pt/ZSM-5 catalyst samples is hydroisomerization (Figure 7). The Pt/ZSM-5-WB-1 sample, which has a higher crystal dispersion, shows the highest selectivity for hexadecane isomers. At 220 °C, the selectivities for hexadecane isomer formation on the Pt/ZSM-5-BD, Pt/ZSM-5-WB-2, and Pt/ZSM-5-WB-1 samples were 13.6%, 14.8%, and 15.7%, respectively.

The low activity in the hydroisomerization reaction compared to hydrocracking is due to the peculiarities of the structure of zeolite ZSM-5 [18]. The three-dimensional channel microporous structure of this zeolite allows the conversion of n-paraffins into highly branched isomers. The isomers can experience stronger diffusion difficulties in the micropores formed by the 10R rings. This leads to their further destruction. The formation of transport pores leads to improved diffusion. Although, with increasing mesopore volume (Table 3), an almost indistinguishable increase in selectivity for hexadecane isomers is observed in samples of the Pt/ZSM-5-BD, Pt/ZSM-5-WB-2, and Pt/ZSM-5-WB-1 series.

Table 5. Yield of n-hexadecane hydrocracking products on Pt/ZSM-5 samples and the ratio of normal and isoparaffins with a carbon chain length between 5 and 13 carbon atoms (i/n).

Sample	Temperature, °C	Yield, wt. %			i/n
		C_2–C_4	C_5–C_{10}	C_{11}–C_{13}	
Pt/ZSM-5-WB-1	220	13.8	32.7	3.6	0.9
	240	29.7	64.1	2.2	1.2
	260	34.1	62.2	1.2	1.8
Pt/ZSM-5-WB-2	220	11.5	28.1	3.2	0.7
	240	31.6	60.5	1.9	1.0
	260	35.5	59.9	1.0	1.5
Pt/ZSM-5-BD	220	8.4	11.9	2.6	0.4
	240	32.3	54.0	1.5	0.8
	260	37.9	56.6	0.8	1.3

4. Conclusions

The physicochemical properties of the products of crystallization of granules (containing 60% by weight of powdered ZSM-5 and 40% by weight of a temporary binder) into granular zeolite using the "less-bind" technology were studied using XRD, XRF, SEM, N_2 adsorption-desorption, mercury porosimetry, and IR spectroscopy with pyridine adsorption.

Two amorphous aluminosilicates with a SiO_2/Al_2O_3 molar ratio of 50 were used as temporary binders. One was prepared from tetraethyl orthosilicate and aluminum nitrate by a two-step sol-gel process, while the other was prepared from aqueous solutions of sodium silicate and aluminum sulfate.

It was found that during the crystallization process, granules prepared using amorphous aluminosilicates as a temporary binder yielded granular ZSM-5 zeolites (d = 1.5 mm, l = 5 mm) with high phase purity and a developed secondary porous structure. The specific surface area of these granules was found to be between 330 and 350 m^2/g, with mesopore volumes between 0.20 and 0.28 cm^3/g and macropore volumes between 0.20 and 0.30 cm^3/g.

It has been found that during the crystallization of granules containing amorphous aluminosilicate with a larger specific surface area and smaller particle size, nanoscale crystals of ZSM-5 zeolite are formed, which are smaller in size.

It has been shown that granular ZSM-5 zeolites with high phase purity and a developed secondary porous structure outperform granular ZSM-5 zeolites with an aluminum oxide binder in terms of mechanical strength.

It was found that a bifunctional catalyst is formed on the H-form of granular ZSM-5 zeolite with a high degree of crystallinity and a developed secondary porous structure when 0.5 wt. % Pt is added. This catalyst exhibits higher activity in the hydroconversion of n-hexadecane (particularly hydrocracking) compared to a catalyst prepared using granular ZSM-5 zeolite with a binder.

Author Contributions: Conceptualization, M.R.A.; Methodology, M.R.A., O.S.T. and B.I.K.; Validation, M.R.A. and O.S.T.; Investigation, M.R.A., O.S.T., N.A.F., A.I.M., D.V.S., A.K.I. and R.A.Z.; Writing—Original Draft Preparation, M.R.A. and N.A.F.; Writing—Review and Editing, M.R.A., N.A.F., O.S.T. and D.V.S.; Resources, A.K.I., R.A.Z., N.A.F. and D.V.S.; Visualization, N.A.F.; Supervision, M.R.A. and B.I.K.; Funding Acquisition, B.I.K. and M.R.A. All authors have read and agreed to the published version of the manuscript.

Funding: This study was performed under a state task at the Institute of Petrochemistry and Catalysis UFRS RAS (project no. FMRS-2024-0012 «Bifunctional Molecular-Sieve Catalytic Systems for the Production of Low-Pour-Point Diesel Fuels»).

Institutional Review Board Statement: Not applicable.

Informed Consent Statement: Not applicable.

Data Availability Statement: The original contributions presented in the study are included in the article, and further inquiries can be directed to the corresponding authors.

Conflicts of Interest: The authors declare no conflicts of interest.

References

1. Primo, A.; Garcia, H. Zeolites as catalysts in oil refining. *Chem. Soc. Rev.* **2014**, *43*, 7548–7561. [CrossRef] [PubMed]
2. Vermeiren, W.; Gilson, J.P. Impact of zeolites on the petroleum and petrochemical industry. *Top. Catal.* **2009**, *52*, 1131–1161. [CrossRef]
3. Database of Zeolite Structure. Available online: http://www.iza-structure.org/databases (accessed on 20 January 2024).
4. Konnov, S.V.; Dubray, F.; Clatworthy, E.B.; Kouvatas, C.; Gilson, J.-P.; Dath, J.-P.; Minoux, D.; Aquino, C.; Valtchev, V.; Moldovan, S.; et al. Novel Strategy for the Synthesis of Ultra-Stable Single-Site Mo-ZSM-5 Zeolite Nanocrystals. *Angew. Chem.* **2020**, *132*, 19721–19728. [CrossRef]
5. Meng, L.; Zhu, X.; Hensen, J.M. Stable Fe/ZSM-5 Nanosheet Zeolite Catalysts for the Oxidation of Benzene to Phenol. *ACS Catal.* **2017**, *7*, 2709–2719. [CrossRef]
6. Jia, C.; Zong, L.; Wen, Y.; Xu, H.; Wei, H.; Wang, X. Synthesis and scale-up of ZSM-5 aggregates with hierarchical structure. *Res. Chem. Intermed.* **2019**, *45*, 3913–3927. [CrossRef]
7. Jia, Y.; Shi, Q.; Wang, J.; Ding, C.; Zhang, K. Synthesis, characterization, and catalytic application of hierarchical nano-ZSM-5 zeolite. *RSC Adv.* **2020**, *10*, 29618–29626. [CrossRef]
8. Zhang, C.; Fan, K.; Ma, G.; Lei, C.; Xu, W.; Jiang, J.; Sun, B.; Zhang, H.; Zhu, Y.; Wen, S. Efficient Synthesis of Mesoporous Nano ZSM-5 Zeolite Crystals without a Mesoscale Template. *Crystals* **2021**, *11*, 1247. [CrossRef]
9. Zheng, Y.; Ning, W.; Wang, Q.; Wei, X.; Li, X.; Pan, M. Hierarchical ZSM-5 zeolite using amino acid as template: Avoiding phase separation and fabricating an ultra-small mesoporous structure. *Microporous Mesoporous Mater.* **2023**, *355*, 112578. [CrossRef]
10. Shen, Y.; Li, H.; Zhang, X.; Wang, X.; Lv, G. The diquaternary ammonium surfactant-directed synthesis of single-unit-cell nanowires of ZSM-5 zeolite. *Nanoscale* **2020**, *12*, 5824–5828. [CrossRef]
11. Wang, P.; Xiao, X.; Pan, Y.; Zhao, Z.; Jiang, G.; Zhang, Z.; Meng, F.; Li, Y.; Fan, X.; Kong, L. Facile Synthesis of Nanosheet-Stacked Hierarchical ZSM-5 Zeolite for Efficient Catalytic Cracking of n-Octane to Produce Light Olefins. *Catalysts* **2022**, *12*, 351. [CrossRef]
12. Tajuddin, N.A.; Kamal, N.A.; Rhymme, N.A.Q.; Sani, S.F. Hierarchical Zeolite Zsm-5 Framework On The Synthesis And Characterization For Catalytic Cracking Of Fluid: A Mini Review. *Malays. J. Anal. Sci.* **2023**, *27*, 499–509.
13. Hao, J.; Zhou, J.; Wang, Y.; Li, L.; Sheng, Z.; Teng, J.; Xie, Z. Pore mouth catalysis promoting n-hexane hydroisomerization over a Pt/ZSM-5 bifunctional catalyst. *Chem. Catal.* **2024**, *4*, 101041. [CrossRef]
14. Yang, N.; Fu, T.; Cao, C.; Wu, X.; Zheng, H.; Li, Z. Constructing hierarchical ZSM-5 coated with small ZSM-5 crystals via oriented-attachment and in situ assembly for methanol-to-aromatics reaction. *Front. Chem. Sci. Eng.* **2024**, *18*, 76. [CrossRef]
15. He, X.; Tian, Y.; Qiao, C.; Liu, G. Acid-driven architecture of hierarchical porous ZSM–5 with high acidic quantity and its catalytic cracking performance. *Chem. Eng. J.* **2023**, *473*, 145334. [CrossRef]
16. Perego, C.; Calemma, V.; Pollesel, P. Naphtha Reforming and Upgrading of Diesel Fractions. In *Zeolites and Catalysis*; Čejka, J.A., Corma, A., Zones, S., Eds.; Wiley-VCH: Weinheim, Germany, 2010; Volume 2, pp. 585–621. [CrossRef]
17. Aljajan, Y.; Stytsenko, V.; Rubtsova, M.; Glotov, A. Hydroisomerization Catalysts for High-Quality Diesel Fuel Production. *Catalysts* **2023**, *13*, 1363. [CrossRef]
18. Soualah, A.; Lemberton, J.L.; Pinard, L.; Chater, M.; Magnoux, P.; Moljord, K. Hydroisomerization of long-chain n-alkanes on bifunctional Pt/zeolite catalysts: Effect of the zeolite structure on the product selectivity and on the reaction mechanism. *Appl. Catal. A* **2008**, *336*, 23–28. [CrossRef]
19. Lee, S.-W.; Ihm, S.-K. Characteristics of Magnesium-Promoted Pt/ZSM-23 Catalyst for the Hydroisomerization of n-Hexadecane. *Ind. Eng. Chem. Res.* **2013**, *52*, 15359–15365. [CrossRef]
20. Romero, D.; Rohling, R.; Meng, L.; Rigutto, M.; Hensen, E.J.M. Shape selectivity in linear paraffins hydroconversion in 10-membered-ring pore zeolites. *J. Catal.* **2021**, *394*, 284–298. [CrossRef]

21. Asgar Pour, Z.; Abduljawad, M.M.; Alassmy, Y.A.; Cardon, L.; Van Steenberge, P.H.M.; Sebakhy, K.O. A Comparative Review of Binder-Containing Extrusion and Alternative Shaping Techniques for Structuring of Zeolites into Different Geometrical Bodies. *Catalysts* **2023**, *13*, 656. [CrossRef]
22. Bingre, R.; Louis, B.; Nguyen, P. An Overview on Zeolite Shaping Technology and Solutions to Overcome Diffusion Limitations. *Catalysts* **2018**, *8*, 163. [CrossRef]
23. Fakin, T.; Risti'c, A.; Mavrodinova, V.; Logar, N.Z. Highly crystalline binder-free ZSM-5 granules preparation. *Microporous Mesoporous Mater.* **2015**, *213*, 108–117. [CrossRef]
24. Travkina, O.S.; Agliullin, M.R.; Filippova, N.A.; Khazipova, A.N.; Danilova, I.G.; Grigor'eva, N.G.; Narenderd, N.; Pavlov, M.L.; Kutepov, B.I. Template-free synthesis of high degree crystallinity zeolite Y with micro–meso–macroporous structure. *RSC Adv.* **2017**, *7*, 32581–32590. [CrossRef]
25. Agliullin, M.R.; Danilova, I.G.; Faizullin, A.V.; Amarantov, S.V.; Bubennov, S.V.; Prosochkina, T.R.; Grigor'eva, N.G.; Paukshtis, E.A.; Kutepov, B.I. Sol-gel synthesis of mesoporous aluminosilicates with a narrow pore size distribution and catalytic activity thereof in the oligomerization of dec-1-ene. *Microporous Mesoporous Mater.* **2016**, *230*, 118–127. [CrossRef]
26. Kachala, V.V.; Khemchyan, L.L.; Kashin, A.S.; Orlov, N.V.; Grachev, A.A.; Zalesskiy, S.S.; Ananikov, V.P. Target-oriented analysis of gaseous, liquid and solid chemical systems by mass spectrometry, nuclear magnetic resonance spectroscopy and electron microscopy. *Russ. Chem. Rev.* **2013**, *82*, 648–685. [CrossRef]
27. Tamura, M.; Shimizu, K.-I.; Satsuma, A. Comprehensive IR study on acid/base properties of metal oxides. *Appl. Catal. A Gen.* **2012**, *433*, 135–145. [CrossRef]
28. Li, S.; Li, J.; Dong, M.; Fan, S.; Zhao, T.; Wang, J.; Fan, W. Strategies to control zeolite particle morphology. *Chem. Soc. Rev.* **2019**, *48*, 885–907. [CrossRef]

MDPI AG
Grosspeteranlage 5
4052 Basel
Switzerland
Tel.: +41 61 683 77 34

Surfaces Editorial Office
E-mail: surfaces@mdpi.com
www.mdpi.com/journal/surfaces

www.ingramcontent.com/pod-product-compliance
Lightning Source LLC
Chambersburg PA
CBHW062057220926
51850CB00010B/2803
* 9 7 8 3 7 2 5 8 5 3 4 9 6 *